Ionic Liquid Crystals

Ionic Liquid Crystals

Special Issue Editor

Giacomo Saielli

MDPI • Basel • Beijing • Wuhan • Barcelona • Belgrade

MDPI

Special Issue Editor
Giacomo Saielli
National Research Council and University of Padua
Italy

Editorial Office
MDPI
St. Alban-Anlage 66
4052 Basel, Switzerland

This is a reprint of articles from the Special Issue published online in the open access journal *Crystals* (ISSN 2073-4352) in 2019 (available at: https://www.mdpi.com/journal/crystals/special_issues/ Ionic_Liquid-Crystals)

For citation purposes, cite each article independently as indicated on the article page online and as indicated below:

LastName, A.A.; LastName, B.B.; LastName, C.C. Article Title. *Journal Name* **Year**, *Article Number*, Page Range.

ISBN 978-3-03921-086-2 (Pbk)
ISBN 978-3-03921-087-9 (PDF)

Contents

About the Special Issue Editor

Giacomo Saielli, Dr., Ph.D., is a researcher of the CNR Institute on Membrane Technology, Padova, Italy. He obtained his Ph.D. in Chemistry at the University of Padova, after graduating in Chemistry at the University of Florence. After the Ph.D., he spent two years as a postdoc at the University of Southampton, UK, and later obtained a short term JSPS fellowship at AIST-Tsukuba (Japan). He has been a Visiting Researcher at The Scripps Research Institute—CA (2010) and at the Lawrence Berkeley National Laboratory—CA (2013), thanks to two STM grants from CNR. He has received the JGA award of the RSC-UK twice, sponsoring a visit at the University of Chicago (2012) and Osaka Sangyo University (2015), and he received the PIFI 2017 (President's International Fellowship Initiative) award as a Visiting Scientist at the CAS Institute of Theoretical Physics in Beijing. His research is focused on computational studies of ionic liquid phases and computational spectroscopy.

Preface to "Ionic Liquid Crystals"

As the Guest Editor, I am delighted to introduce this book edition of the Special Issue on Ionic Liquid Crystals (ILCs) published by the journal Crystals. I was prompt to accept the invitation as Guest Editor for the Special Issue by an awareness of the peculiarity of ILCs: Despite the fact that they can be seen as both ionic liquids and/or liquid crystals, the combination of the properties of these latter two materials into a single substance promotes the emergence of new properties and novel features. Obviously, new issues and difficulties also appear, which necessitate a renewed effort, both from an experimental and theoretical point of view, in order to better understand their behavior, the structure-properties relationships, and to exploit the possible technological applications.

Ionic liquid crystals are attracting more and more attention in the literature and there is a rapid increase in the number of papers dealing with ILCs. The detailed understanding of their properties is, however, still far from being complete and, although ILCs merge together the positive characteristics of both ionic liquids and liquid crystals, they also combine their drawbacks, such as, for example, a relatively high viscosity. The design of novel ionic liquid crystalline phases with a lower viscosity and better performance is indeed a very difficult task.

For this reason, this Special Issue collects several papers from authors belonging to diverse disciplines—engineering, synthetic organic chemistry, optical and magnetic spectroscopy, theoretical physics, computational chemistry—reflecting the wide scope of the field and the vast array of techniques needed to investigate ILCs. It has been a pleasure and an honor to receive the submissions from many esteemed authors, colleagues, and friends, and I wish to thank them for their contribution. I am also extremely pleased to say that all papers were very well received from the many reviewers selected from an international panel of expert in the field.

Moreover, I wish to thank the Editorial Office of the journal Crystals, particularly Mr. Adonis Tao, for inviting me to guest edit the Special Issue and for the help during the whole process.

I hope this book will further stimulate work in the field of ILCs and, more importantly, I hope it will highlight the need for an interdisciplinary approach to their study.

Giacomo Saielli
Special Issue Editor

crystals

MDPI

Editorial

Special Issue Editorial: Ionic Liquid Crystals

Giacomo Saielli

CNR Institute on Membrane Technology, Unit of Padova, and Department of Chemical Sciences, University of Padova, via Marzolo, 1 - 35131 Padova, Italy; giacomo.saielli@unipd.it

Received: 23 May 2019; Accepted: 26 May 2019; Published: 27 May 2019

The term "Ionic Liquid Crystals" (ILCs) clearly results from the blending of the well-known "Ionic Liquids" (ILs) and "Liquid Crystals" (LCs) names of the corresponding materials. The concatenating word *Liquid* is crucial since this is the property that makes all three types of materials so important: Without the key feature of being fluid, there would be not such notable interest in the phase behavior of either an ionic or molecular solid. Coincidentally, both ILs and LCs were discovered in 1888 [1,2] and they remained just an academic curiosity for many decades, until industrial applications eventually took off. This happened during the 1970s for liquid crystals, after the synthesis of a new family of LCs based on cyanobiphenyls, stable to oxidation and light irradiation [3]; similarly, ILs had to wait until the discovery of air and water stable imidazolium salts in the 1990s [4] before they started to become appealing for industrial processes.

Thermotropic Ionic Liquid Crystals were first officially reported some 50 years later [5], compared to the "parent" compounds, but in fact they have been known since ancient times, since soaps, that is metal alkanoates, exhibit ionic liquid crystal phases. ILCs can be viewed as ILs that, at some intermediate temperature between the isotropic liquid and the crystal phase, also exhibit a liquid-crystalline (LC) mesophase. Most of the ILC compounds known today are composed by the same type of cations and anions usually found in ILs; however, because of the presence of relatively long carbon chains, micro-segregation leads to the formation of LC phases, almost invariably of smectic type, that is layered. Other LC phases encountered in ionic systems are columnar and cubic ones. In contrast, the ionic nematic phase is extremely rare and the quest for a family of compounds showing a stable ionic nematic phase, near room temperature and with a relatively large thermal range of stability, is an active field of research. In any case, even for the most common kinds of ILCs, the detailed understanding of the relationship between the molecular structure of cation and anion and the phase behavior and thermal stability is far from being understood.

The potential applications of ILCs span a wide range of options and they have been tested in several proof-of-principle devices. The special solvation properties of ILs, combined with the partial orientational and/or translational order of LCs, make ILCs promising media in all cases where a transport of mass and/or charge is needed. There is a main drawback, though, that is the relatively high viscosity. This is the reason why deeper and more thorough investigations of ILCs are necessary, in order to understand how the many details of the molecular structure of cations and anions affect the macroscopic properties and the thermal range and type of mesophases of ILCs. These, in fact, depend on the interplay of a number of steric, van der Waals, H-bonding, and electrostatic interactions and their modeling is an arduous task.

This Special Issue on Ionic Liquid Crystals aims at gathering together some of the specialists working with ILCs, to shed light on the properties and behavior of ILCs. The papers cover many aspects of ILCs science and technology from organic to computational chemistry, from physical chemistry to engineering applications, thus reflecting the many interests of the community of scientists active in the field.

In Reference [6], Liu et al. have investigated the ability of nanoparticles to trap ionic impurities in a LC cell and therefore the possibility to reduce the residual DC current. The residual direct current voltage caused by the accumulation of mobile ions is here prohibited. Although the system investigated is not stricto sensu a thermotropic ILC, the work clearly highlights the importance of the precise control and understanding of ionic interaction in LC mesophase for real applications.

In Reference [7], Bhowmik and co-workers report the synthesis and characterization of a series of viologen-based ionic liquid crystals having 4-*n*-alkylbenzenesulfonates as counter-anions. Viologens have interesting redox and electrochromic properties, therefore the investigation of their mesophase behavior is of utmost importance in view of possible applications.

In Reference [8], Laschat, Giesselmann and co-workers report the synthesis and characterization of discotic ILCs based on crown ethers. The systems formed columnar mesophases and the authors observed an improved electronic transport, namely the hole mobility, in macroscopically aligned thin films. They excluded the presence of channels for fast cation transport; rather they found that the ion migration is dominated by non-coordinating anions propagating trough the ordered medium.

An interesting investigation of the relationship between molecular structure of the constituent cations and anions and the phase behavior of the material, is reported by Goossens et al. in Reference [9]. They prepared a series of 4,5-bis(*n*-alkyl)azolium salts and studied their behavior. The authors observed that the presence of substituents on the 4- and 5-positions of the imidazolium ring increases the melting points and lowers the clearing points compared to the 1,3-disubtituted analogues.

An entirely different perspective on ILCs is presented by Cao and Wang. In their paper, Reference [10], they investigated imidazolium salts using fully atomistic molecular dynamics (MD) simulations. They prepared different crystal structures of 1-tetradecyl-3-methylimidazolium nitrate and heated them up to the transition into the smectic phase. They observed that all systems melt into the same SmA phase. The systems go through a metastable state which is characterized by an orientation of the chains almost perpendicular to the smectic layers. The power of MD simulations is therefore highlighted by the possibility to study phases not accessible by experiments and to rationalize their stability.

Nuclear Magnetic Resonance spectroscopy is a fundamental experimental technique to investigate ordered phases; measuring ^{13}C-^{1}H dipolar couplings of samples in an orientationally ordered medium allows to obtain orientational order parameters of the corresponding C-H bonds. Based on this technique, Dvinskikh and co-worker in Reference [11] have analyzed the orientational order of the thermotropic ILC 1-tetradecyl-3-methylimidazolium nitrate in the thermal range of stability of the smectic phase. They reported a significantly lower value of the orientational order parameters compared to conventional non-ionic LC phases.

Finally, we also presented a contribution concerned with MD simulations, using highly coarse-grained models, of ILCs [12]. We considered a mixture of ellipsoidal particles based on the Gay–Berne potential, positively charged to represent the cations, and spherical Lennard–Jones particles negatively charged to represent the anions. Though extremely simplified, the investigation of the phase diagram of such a model system showed the appearance of a very stable ionic nematic phase between the isotropic phase and the smectic phase.

To conclude, I believe that this Special Issue on Ionic Liquid Crystals touches on the latest advancements in several aspects related to ILCs science: Synthesis of novel compounds, spectroscopic studies, MD simulations, and investigations of both structural and dynamic properties. I wish to express my deepest and sincere gratitude to all authors who contributed, for having submitted manuscripts of such excellent quality. I also wish to thank the Editorial Office of *Crystals* for the fast and professional handling of the manuscripts during the whole submission process and for the help provided.

Conflicts of Interest: The authors declare no conflict of interest.

References

1. Gabriel, S.; Weiner, J. Ueber einige Abkömmlinge des Propylamins. *Ber. Dtsch. Chem. Ges.* **1888**, *21*, 2669–2679. [CrossRef]
2. Reinitzer, F. Beiträge zur Kenntniss des Cholesterins. *Monatsh. für Chem. (Wien)* **1888**, *9*, 421–441. [CrossRef]
3. Gray, G.W.; Harrison, K.J.; Nash, J.A. New family of nematic liquid crystals for displays. *Electr. Lett.* **1973**, *9*, 130–131. [CrossRef]
4. Wilkes, J.S.; Zaworotko, M.J. Air and Water Stable 1-Ethyl-3-methylimidazolium Based Ionic Liquids. *J. Chem. Soc. Chemm. Commun.* **1992**, 965–967. [CrossRef]
5. Knight, G.A.; Shaw, B.D. Long-Chain Alkylpyridines and Their Derivatives. New Examples of Liquid Crystals. *J. Chem. Soc.* **1938**, 682–683. [CrossRef]
6. Liu, Y.; Sang, J.; Liu, H.; Xu, H.; Zhao, S.; Sun, J.; Lee, J.H.; Jeong, H.-C.; Seo, D.-S. Decreasing the Residual DC Voltage by Neutralizing the Charged Mobile Ions in Liquid Crystals. *Crystals* **2019**, *9*, 181. [CrossRef]
7. Bhowmik, P.K.; Chang, A.; Kim, J.; Dizon, E.J.; Principe, R.C.G.; Han, H. Thermotropic Liquid-Crystalline Properties of Viologens Containing 4-n-alkylbenzenesulfonates. *Crystals* **2019**, *9*, 77. [CrossRef]
8. Staffeld, P.; Kaller, M.; Ehni, P.; Ebert, M.; Laschat, S.; Giesselmann, F. Improved Electronic Transport in Ion Complexes of Crown Ether Based Columnar Liquid Crystals. *Crystals* **2019**, *9*, 74. [CrossRef]
9. Goossens, K.; Rakers, L.; Shin, T.J.; Honeker, R.; Bielawski, C.W.; Glorius, F. Substituted Azolium Disposition: Examining the Effects of Alkyl Placement on Thermal Properties. *Crystals* **2019**, *9*, 34. [CrossRef]
10. Cao, W.; Wang, Y. Phase Behaviors of Ionic Liquids Heating from Different Crystal Polymorphs toward the Same Smectic-A Ionic Liquid Crystal by Molecular Dynamics Simulation. *Crystals* **2019**, *9*, 26. [CrossRef]
11. Dai, J.; Kharkov, B.B.; Dvinskikh, S.V. Molecular and Segmental Orientational Order in a Smectic Mesophase of a Thermotropic Ionic Liquid Crystal. *Crystals* **2019**, *9*, 18. [CrossRef]
12. Margola, T.; Satoh, K.; Saielli, G. Comparison of the Mesomorphic Behaviour of 1:1 and 1:2 Mixtures of Charged Gay-Berne GB(4.4,20.0,1,1) and Lennard-Jones Particles. *Crystals* **2018**, *8*, 371. [CrossRef]

crystals

MDPI

Article

Decreasing the Residual DC Voltage by Neutralizing the Charged Mobile Ions in Liquid Crystals

Yang Liu [1,2,*], Jingxin Sang [1,2], Hao Liu [1,2], Haiqin Xu [1,2], Shuguang Zhao [1,2], Jiatong Sun [1,2,*], Ju Hwan Lee [3], Hae-Chang Jeong [3] and Dae-Shik Seo [3,*]

[1] College of Information Science and Technology, Donghua Uiversity, 2999 North Renmin Road, Songjiang District, Shanghai 201620, China; 2161523@mail.dhu.edu.cn (J.S.); liuhao@dhu.edu.cn (H.L.); xuhaiqin@dhu.edu.cn (H.X.); sgzhao@dhu.edu.cn (S.Z.)

[2] Engineering Research Center of Digitized Textile & Fashion Technology, Ministry of Education, Donghua University, 2999 North Renmin Road, Songjiang District, Shanghai 201620, China

[3] Information Display Device Laboratory, Department of Electrical and Electronic Engineering, Yonsei University, 50 Yonsei-ro, Seodaemun-gu, Seoul 120-749, Korea; whitewing23@yonsei.ac.kr (J.H.L.); gundamhc@yonsei.ac.kr (H.-C.J.)

* Correspondence: liuyang@dhu.edu.cn (Y.L.); jsun@dhu.edu.cn (J.S.); dsseo@yonsei.ac.kr (D.-S.S.); Tel.: +86-021-6779-2135 (Y.L. & J.S.); +82-02-2123-7727 (D.-S.S.)

Received: 14 March 2019; Accepted: 25 March 2019; Published: 27 March 2019

Abstract: The decrease of the residual direct current (DC) voltage (V_{rdc}) of the anti-parallel liquid crystal (LC) cell using silver (Ag)-doped Polyimide (Ag-d-PI) alignment layers is presented in this manuscript. A series of Ag/PI composite thin layers are prepared by spurting or doping PI thin layers with Ag nano-particles, and Ag/PI composite thin layers are highly transparent and resistive. LC are homogeneously aligned between 2.0 mg/mL Ag-d-PI alignment layers, and the V_{rdc} of the cell that assembled with Ag-d-PI alignment layers decreases about 82%. The decrease of V_{rdc} is attributed to the trapping and neutralizing of mobile ions by Ag nano-particles. Regardless of the effect of Ag nano-particles on the conductivity of Ag-d-PI alignment layers, the voltage holding ratio (VHR) of the cells is maintained surprisingly. The experiment results reveal a simple design for a low V_{rdc} LC cell.

Keywords: liquid crystal; alignment layer; residual DC; Ag nano-particles doping

---

1. Introduction

Liquid crystals (LC) are widely used in electro-optic devices because of their unique electro-optic anisotropy; however, the mobile ions in LC cause a lot of problems relating to LC switching. The moving of mobile ions driven by electric forces towards alignment layers results in their accumulation on alignment layers, which finally generates residual direct current (DC) voltage (V_{rdc}) inside LC cells and adversely affects LC' switching [1–6]. During the last several decades, a series of researches focused on distinguishing, detecting mobiles ions, and revealing the influences of mobile ions shifting on LC switching were conducted, and nowadays a lot of explorations are carried out to reduce mobile ions' adverse functions on LC [7–12]

A lot of attempts have been adopted to prevent the influences of mobile ions on LC electro-optical performances, such as designing special LC molecules, purifying LC, doping LC [13,14], replacing the polyimide (PI) alignment layers with conductive materials [15–18], and photo-aligning LC [19–21], etc. Compared with other methods, doping is much easier; however, doping LC with nano-materials brings new issues, for instance, the doped nano-materials are too poor to be dispersed, and the aggregation of these nano-materials makes LC insensitively respond to external voltage. The aggregation of nano-materials is partially prevented by tightly limiting the amount of doped nano-materials; however, because of the electric field, the doped nano-materials in LC move towards alignment layers and are accumulated on alignment layers, which enhances V_{rdc} generation. Replacing PI alignment layers with conductive alignment layers significantly reduces V_{rdc} on cells; however, the conductive alignment layers in the cells raise the issue of a voltage holding ratio (VHR) decrease [22,23].

Micro silver (Ag) particles are highly transparent and conductive and have been adopted to accelerate LC optical switching and trap the ionic charges. In this manuscript, Ag-spurted PI (Ag-s-PI) alignment layers and Ag nano-particles-doped PI (Ag-d-PI) alignment layers are prepared and used to trap the mobile ions in LC, and the residual DC of the cell assembled with Ag-d-PI alignment layers decreases obviously. As shown in Figure 1, the displacement polarization occurs in Ag nano-particles, when the external voltage is applied on the cell that assembled with Ag-d-PI composite alignment layers. The Ag nano-particles are immobilized by PI molecules, which restricts their shift to the LC medium. The mobile ions driven by electric forces move towards and gather near Ag-d-PI composite alignment layers, and the positive and negative charges carried by mobile ions are trapped and neutralized by Ag nano-particles. In this case, the V_{rdc} caused by the accumulation of mobile ions is prohibited. Because the amount of doped Ag nano-particles is limited up to 2 mg/mL (m_{Ag}/V_{PI}), the electrical conductivity change of Ag-d-PI composite alignment layers could be ignored, and the decrease of voltage holding ratio on the cell is prevented.

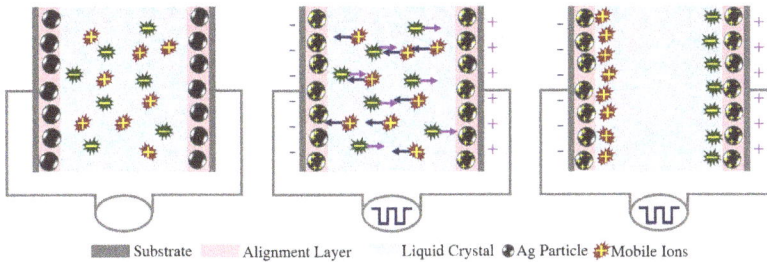

Figure 1. The schematic of mobile ions accumulating on Ag-d-PI composite alignment layers.

2. Materials and Methods

Ag-doped PI solutions were prepared by doping Ag nano-particles (particle size < 100 nm, Sigma-Aldrich) into homogeneous PI solutions (SE7792, Nissan Chemical Corporation) with their concentrations maintained at 0.2 mg/mL, 0.5 mg/mL, 1.0 mg/mL, and 2.0 mg/mL, respectively, and Ag/PI solution was sonicated at room temperature for 30 min to disperse Ag nano-particles uniformly. Ag-d-PI thin layers were prepared by spin-coating the prepared Ag/PI solutions on ITO substrates, and Ag nano-particles spurted PI thin layers were prepared by spurting Ag nano-particle solutions (Ag/acetone, 0.2 mg/mL, 0.5 mg/mL, 1.0 mg/mL, and 2.0 mg/mL) onto spin-coated PI alignment layers. Considering that the rubbing process is necessary to align LC, and during the rubbing process Ag nano-particles may be partially removed, two Ag nano-particles-spurted PI alignment layers were prepared. One is spurting Ag nano-particles on the PI alignment layers and then followed with the rubbing process (Ag-s-PI), and the other is spurting Ag nano-particles on the rubbed PI alignment layers (Ag-s-rPI). The transmittance spectra of Ag/PI thin layers on glass slides were characterized by using a double-beam UV-Vis spectrophotometer (UV-2101, Shimadzu, Japan) and a 3-D laser-beam profiler system.

Anti-parallel cells with the cell gaps of 60 and 5 μm were assembled, and the commercial LC (n_e = 1.5702, n_0 = 1.4756, and $\Delta\varepsilon$ = 10.7; from Merck) was injected into the fabricated cells. Alignment of LC between Ag/PI composite alignment layers was characterized by a polarized optical microscopy (POM, BXP 51, Olympus); the anchoring energy of LC on Ag/PI composite alignment layers, the V_{rdc} and the capacitance of the cells were evaluated by means of a capacitance-voltage (C-V) hysteresis method (LCR meter, Agilent 4284A) with the maximum bias voltage of 10 V and a step bias voltage of 0.1 V.

3. Results and Discussion

As shown in Figure 2, the prepared Ag/PI alignment layers are transparent and have a transmittance above 82%; no obvious transmittance difference is observed between Ag-s-rPI, Ag-s-PI and Ag-d-PI alignment layers. Besides the transmittance decrease, the aggregation of Ag nano-particles may cause more serious issues, for instance, the aggregated Ag nano-particles block light and result in the non-uniform transparency of thin layers. The blocking performance of Ag/PI composite thin layers was characterized by using a 3D profiler as shown in Figure 3, and no significant difference is observed between the light source and the laser crossing Ag/PI composite thin layers in distribution and intensity, which reveals the potential application of Ag/PI composite thin layers for real LC devices.

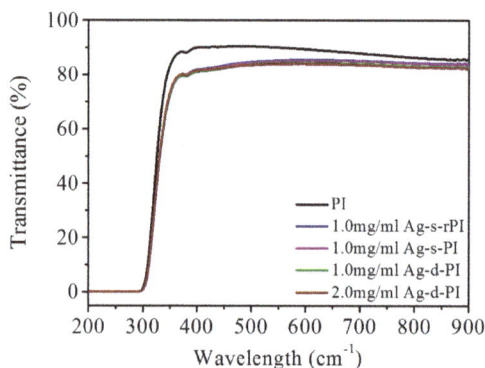

Figure 2. The transmittance of Ag/PI composite alignment layers (Coated on glass substrate).

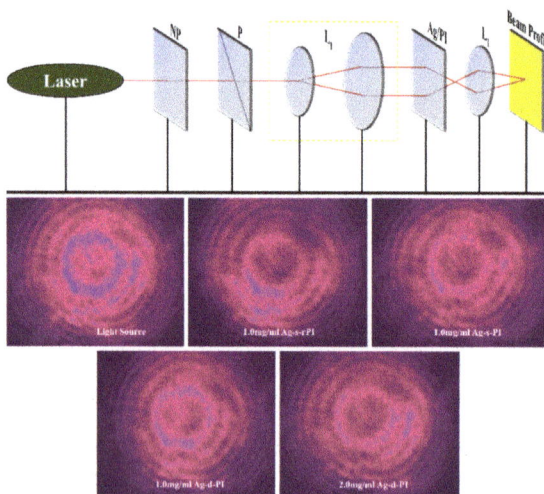

Figure 3. The schematic diagram of 3-D profiler and the captured images of laser-crossed Ag/PI alignment layers.

The alignment of LC between Ag/PI composite alignment layers was confirmed by using POM as shown in Figure 4. Obvious light leakages are observed from the cell assembled with Ag-s-rPI alignment layers, and due to the Ag nano-particles aggregation effect, the light leakages become more serious while increasing the amount of spurted Ag nano-particles. The alignment of LC sandwiched between Ag-s-PI alignment layers is more uniform compared with the mentioned Ag-s-rPI alignment layers, which indicates that the aggregated Ag nano-particles have been removed during the rubbing process. Even the concentrated Ag nano-particles, as high as 2.0 mg/mL, are doped into PI solutions; LC are homogeneously aligned between Ag-d-PI alignment layers and no obvious light leakages are observed.

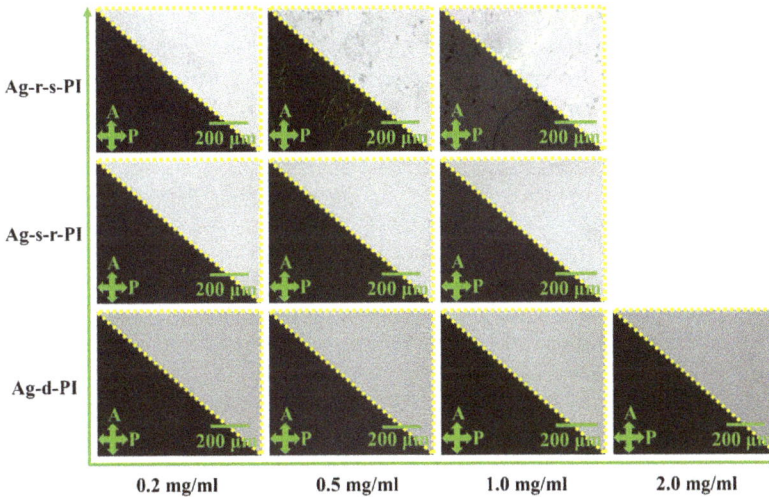

Figure 4. POM images of LC sandwiched between Ag-s-rPI thin layers, Ag-s-PI thin layers and Ag-d-PI thin layers, respectively.

The polar anchoring energy of LC sandwiched between Ag/PI composite alignment layers varies a lot as shown in Figure 5. LC sandwiched between Ag-s-PI alignment layers and Ag-s-rPI alignment layers have similar polar anchoring energies, however, the polar anchoring energy of LC sandwiched between Ag-d-PI composite alignment layers decreases a lot in comparison. The surfaces of Ag nano-particles-spurted PI alignment layers are almost covered with Ag nano-particles, and the surfaces of Ag-d-PI alignment layers are almost PI molecules conversely. Thus, the difference between polar anchoring energies is due to the surface composition alterations by spurting or doping Ag nano-particles, which tunes the interactions between LC and Ag/PI composite thin layers.

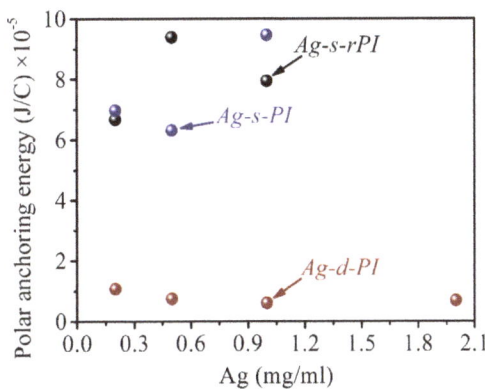

Figure 5. The polar anchoring energy of LC sandwiched between Ag-s-rPI alignment layers, Ag-s-PI alignment layers and Ag-d-PI thin layers, respectively.

When the external voltage is applied on the cell, the mobile ions in LC are driven to shift towards alignment layers and trapped in the localized defect regions, and in this case, V_{rdc} is generated. The fractional coverage of the alignment layer surface, which indicates alignment layers ability to trap mobile ions, is determined as φ_s, and

$$\varphi_s = \frac{\delta}{\delta_s},$$

here, δ and δ_s are the surface density of the adsorption sites occupied by ions and the surface density of all adsorption sites on the alignment layer surface, respectively. After the displacement polarizing of Ag nano-particles, the Ag nano-particles in PI layers trap and neutralize the mobile ions, and thus the φ_s of Ag/PI composite alignment layers get much lower compared with that of the conventional PI alignment layers.

During the rubbing process or driven by external electric filed, partial Ag nano-particles spurted on PI alignment layers are detached and dive into LC. A small amount of detached Ag nano-particles in LC trap and neutralize the charged mobile ions and decrease V_{rdc}. However, if the amount of detached Ag nano-particles is large, the Ag nano-particles shift towards the alignment layers and contribute to the generation of V_{rdc}. By doping and immobilizing Ag nano-particles in PI alignment layers, the V_{rdc} generated by detached Ag nano-particles is prevented. As shown in Table 1 and Figure 6, the V_{rdc} of the cell assembled with Ag-d-PI alignment layers is as low as 0.1132 V when the concentration of doped Ag nano-particles in PI alignment layers is increased to 2.0 mg/mL.

Table 1. V_{rdc} of cells assembled with Ag/PI composite alignment layers.

	Ag-s-PI			Ag-s-rPI			Ag-d-PI		
	V_{rdc}^{+}	V_{rdc}^{-}	V_{rdc}	V_{rdc}^{+}	V_{rdc}^{-}	V_{rdc}	V_{rdc}^{+}	V_{rdc}^{-}	V_{rdc}
0.2	0.6735	0.5899	0.6317	0.7455	0.8417	0.7936	0.4915	0.5107	0.5011
0.5	0.6371	0.6685	0.6528	0.8900	0.8608	0.8754	0.3234	0.3518	0.3376
1.0	-	-	-	-	-	-	0.2121	0.2049	0.2085
2.0	-	-	-	-	-	-	0.1091	0.1173	0.1132

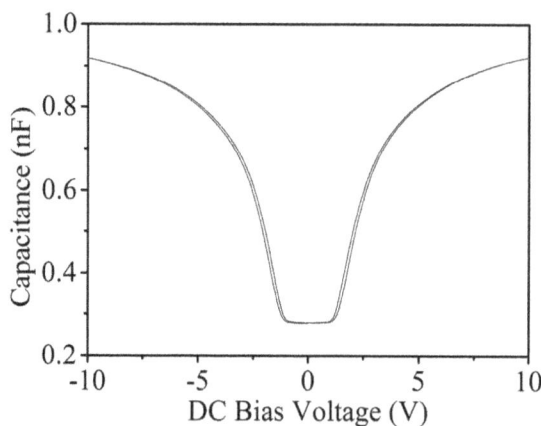

Figure 6. Voltage-dependence capacitance hysteresis characteristic of the cell fabricated from 2.0 mg/mL Ag-d-PI thin layers.

Trapping and neutralizing the charged mobile ions in LC by the displacement polarization in Ag nano-particles may cause the undesired screening effect and the decrease of VHR, and the capacitance of the cells assembled with Ag-d-PI alignment layers is characterized and shown in Figure 7. The capacitance of the cells assembled with Ag-d-PI alignment layers is found slightly decreased with the increase of the amount of doped Ag nano-particles; however, the maximum capacitance of each cell is almost maintained at about 2.4. By increasing the frequency of external voltage on the cells up to 10 khz, a slightly red shift of the capacitance is observed; however, no significant capacitance change in value is observed. The threshold voltage of the cells maintains at about 1.4 V regardless of the increase of the amount of doped Ag nano-particles or the frequency of external voltage, and the maintained capacitance and threshold voltage of cells is attributed to the fact that barely any electrical conductivity change is generated by the different amounts of Ag nano-particles doping.

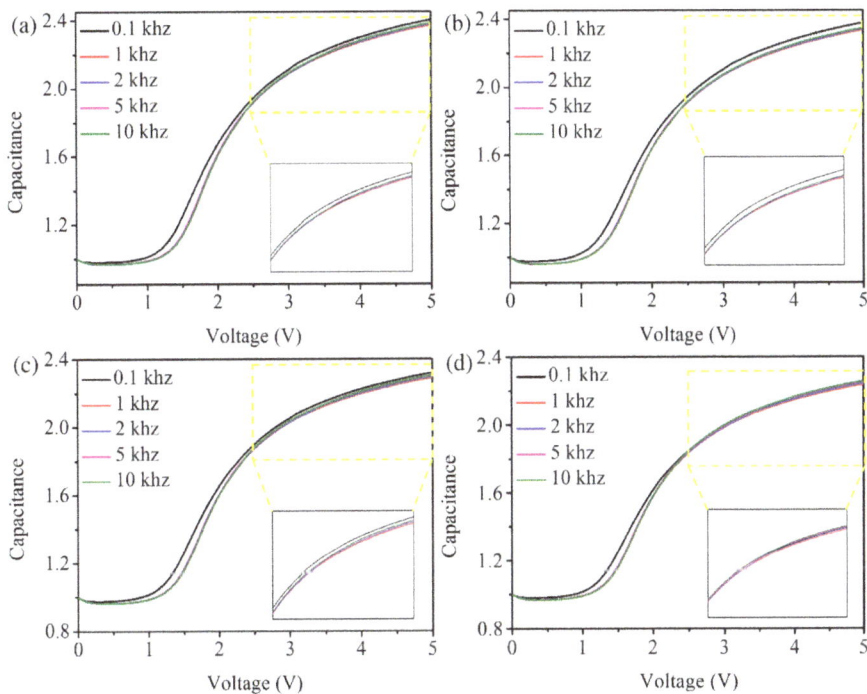

Figure 7. The capacitance-voltage curve of LC sandwiched between (**a**) 0.2 mg/mL, (**b**) 0.5 mg/mL, (**c**) 1.0 mg/mL, and (**d**) 2.0 mg/mL Ag-d-PI composite thin layers.

4. Conclusions

In conclusion, LC is homogenously aligned between Ag-d-PI alignment layers, and the mobile ions in LC are trapped and neutralized by Ag nano-particles due to their displacement polarization when the external voltage is on, which decreases the V_{rdc} on cells effectively. Compared with the cells assembled with conductive alignment layer cells, the VHR of the cells assembled with Ag-d-PI composite alignment layers is maintained. The extremely simple design adopted to deduce the V_{rdc} on the cells in this manuscript is worth more attentions.

Author Contributions: Y.L. and J.S. (Jiatong Sun) conceived the original idea and wrote the manuscript; Y.L., J.S. (Jiatong Sun), J.S. (Jingxin Sang), J.H.L. and H.-C.J. performed the experiments, H.L., H.X., S.Z. and D.-S.S. analyzed the data. Y.L. supervised and directed the research.

Funding: This work was sponsored by Shanghai Sailing Program (No. 18YF1400900), the National Natural Science Foundation of China (NSFC, No. 6180030581) and Fundamental Research Funds for the Central Universities (No. 2232018D3-29 and No. 2232017D-10).

Conflicts of Interest: The authors declare no conflict of interest.

References

1. Mizusaki, M.; Miyashita, T.; Uchida, T. Generation mechanism of residual direct current voltage in a liquid crystal display and its evaluation parameters related to liquid crystal and alignment layer materials. *J. Appl. Phys.* **2007**, *102*, 014904. [CrossRef]
2. Mizusaki, M.; Miyashita, T.; Uchida, T. Behavior of ion affecting image sticking on liquid crystal displays under application of direct current voltage. *J. Appl. Phys.* **2010**, *108*, 104903. [CrossRef]
3. Mizusaki, M.; Miyashita, T.; Uchida, T. Kinetic analysis of image sticking with adsorption and desorption of ions to a surface of an alignment layer. *J. Appl. Phys.* **2012**, *112*, 044510. [CrossRef]
4. Xu, D.; Peng, F.; Chen, H.; Yuan, J.; Wu, S.-T.; Li, M.-C.; Lee, S.-L.; Tsai, W.-C. Image sticking in liquid crystal displays with lateral electric fields. *J. Appl. Phys.* **2014**, *116*, 193102. [CrossRef]
5. Kim, D.H.; Kim, J.H.; Kwon, Y.R.; Ahn, S.H.; Srivastava, A.K.; Lee, S.H. Investigation on ion movement in the fringe-field switching mode depending on resistivity of alignment layer and dielectric anisotropic sign of liquid crystal. *Liq. Cryst.* **2015**, *42*, 486–491. [CrossRef]
6. Chen, P.-A.; Yang, K.-H. Ionic effects on electro-optics and residual direct current voltages of twisted nematic liquid crystal cells. *Liq. Cryst.* **2018**, *45*, 1032–1039. [CrossRef]
7. Seen, S.M.; Kim, M.S.; Lee, S.H. Image Sticking Resistant Liquid Crystal Display Driven by Fringe Electric Field for Mobile Applications. *Jpn. J. Appl. Phys.* **2010**, *49*, 050208. [CrossRef]
8. Choi, N.; Jung, J.; Cheong, B.; Yoon, H.; Hong, M. Reduction of residual DC voltage via RC matching in LCD. *Mater. Res. Express* **2018**, *5*, 126305. [CrossRef]
9. Gao, L.; Dai, Y.; Li, T.; Tang, Z.; Zhao, X.; Li, Z.; Meng, X.; He, Z.; Li, J.; Cai, M.; Wang, X.; Zhu, J.; Xing, H.; Ye, W. Enhancement of Image Quality in LCD by Doping γ-Fe$_2$O$_3$ Nanoparticles and Reducing Friction Torque Difference. *Nanomaterials* **2018**, *8*, 911. [CrossRef]
10. Inoue, D.; Miyake, T.; Sugimoto, M. A mechanism of short-term image-sticking phenomenon caused by flexoelectric effect in IPS LCD. *IEICE T. Electr.* **2018**, *E101-C*, 846–850. [CrossRef]
11. Ju, C.; Kim, T.; Kang, H. Liquid crystal alignment behaviors on capsaicin substituted polystyrene films. *RSC Adv.* **2017**, *7*, 41376–41383. [CrossRef]
12. Mizusaki, M.; Enomoto, S.; Hara, Y. Generation mechanism of residual direct current voltage for liquid crystal cells with polymer layers produced from monomers. *Liq. Cryst.* **2017**, *44*, 609–617. [CrossRef]
13. Chen, W.-T.; Chen, P.-S.; Chao, C.-Y. Effect of doped insulating nanoparticles on the electro-optical characteristics of nematic liquid crystals. *Jpn. J. Appl. Phys.* **2009**, *48*, 015006. [CrossRef]
14. Lee, H.M.; Chung, H.-K.; Park, H.-G.; Jeong, H.-C.; Han, J.-J.; Cho, M.-J.; Lee, J.-W.; Seo, D.-S. Residual DC voltage-free behaviour of liquid crystal system with nickel nanoparticle dispersion. *Liq. Cryst.* **2014**, *41*, 247–251. [CrossRef]
15. Liu, Y.; Lee, J.H.; Seo, D.-S.; Li, X.-D. Ion-beam-spurted dimethyl-sulfate-doped PEDOT: PSS composite-layer-aligning liquid crystal with low residual direct-current voltage. *Appl. Phys. Lett.* **2016**, *109*, 101901. [CrossRef]
16. Liu, Y.; Zhang, Y.; Oh, B.-Y.; Seo, D.-S.; Li, X.-D. Super-fast switching of liquid crystals sandwiched between highly conductive graphene oxide/dimethyl sulfate doped PEDOT: PSS composite layers. *J. Appl. Phys.* **2016**, *119*, 194505. [CrossRef]
17. Liu, Y.; Park, H.-G.; Lee, J.H.; Seo, D.-S.; Kim, E.-M.; Heo, G.-S. Electro-optical switching of liquid crystals sandwiched between ion-beam-spurted graphene quantum dots-doped PEDOT: PSS composite layers. *Opt. Express* **2015**, *23*, 34071. [CrossRef] [PubMed]
18. Lee, T.R.; Kim, J.H.; Lee, S.H.; Jun, M.C.; Baik, H.K. Investigation on newly designed low resistivity polyimide-type alignment layer for reducing DC image sticking of in-plane switching liquid crystal display. *Liq. Cryst.* **2017**, *44*, 738–747. [CrossRef]
19. Nakanishi, Y.; Hanaoka, K.; Shibasaki, M.; Okamoto, K. Relation between monomer structure and image sticking phenomenon of polymer-sustained-alignment liquid crystal displays. *Jpn.J. Appl. Phys.* **2011**, *50*, 051702. [CrossRef]
20. Lim, Y.J.; Jeong, I.H.; Kang, H.-S.; Kundu, S.; Lee, M.-H.; Lee, S.H. Reduction of the residual DC in the photoaligned twisted nematic liquid crystal display using polymerized reactive mesogen. *Appl. Phys. Express* **2012**, *5*, 081701. [CrossRef]

21. Tseng, M.-C.; Yaroshchuk, O.; Bidna, T.; Srivastava, A.K.; Chigrinov, V.; Kwok, H.-S. Strengthening of liquid crystal photoalignment on azo dye films: passivation by reactive mesogens. *RSC Adv.* **2016**, *6*, 48181–48188. [CrossRef]
22. Jeon, Y.-J.; Hwang, J.-Y.; Seo, D.-S.; Kim, H.-Y. Voltage holding ratio and residual DC property of the IPS-LCD on rubbed polymer layers by voltage-transmittance hysteresis method. *Mol. Cryst. Liq. Cryst.* **2004**, *410*, 369–380. [CrossRef]
23. Liu, Y.; Lee, J.H.; Seo, D.-S. Ion beam fabrication of aluminum-doped zinc oxide layer for high-performance liquid crystals alignment. *Optic. Express* **2016**, *24*, 17424. [CrossRef] [PubMed]

crystals

MDPI

Article

Thermotropic Liquid-Crystalline Properties of Viologens Containing 4-n-alkylbenzenesulfonates [†]

Pradip K. Bhowmik *, Anthony Chang, Jongin Kim, Erenz J. Dizon, Ronald Carlo G. Principe and Haesook Han

Department of Chemistry and Biochemistry, University of Nevada Las Vegas, 4505 S. Maryland Parkway, Box 454003, Las Vegas, NV 89154, USA; changa14@unlv.nevada.edu (A.C.); kimj80@unlv.nevada.edu (J.K.); Dizone2@unlv.nevada.edu (E.J.D.); princr1@unlv.nevada.edu (R.C.G.P.); hanh3@unlv.nevada.edu (H.H.)
* Correspondence: pradip.bhowmik@unlv.edu; Tel.: +1-702-895-0885; Fax: +1-702-895-4072
† This article is dedicated to the memory of Dr. Ananda M. Sarker (1952–2018): deceased on 30 August 2018.

Received: 28 December 2018; Accepted: 28 January 2019; Published: 1 February 2019

Abstract: A series of viologens containing 4-n-alkylbenzenesulfonates were synthesized by the metathesis reaction of 4-n-alkylbenzenesulfonic acids or sodium 4-n-alkylbezenesulfonates with the respective viologen dibromide in alcohols. Their chemical structures were characterized by Fourier Transform Infrared, ^{1}H and [13]C Nuclear Magnetic Resonance spectra and elemental analysis. Their thermotropic liquid-crystalline (LC) properties were examined by differential scanning calorimetry and polarizing optical microscopy. They formed LC phases above their melting transitions and showed isotropic transitions. As expected, all the viologen salts had excellent stabilities in the temperature range of 278–295 °C as determined by thermogravimetric analysis.

Keywords: viologens; 4-n-alkylbenzenesulfonic acids; metathesis reaction; ionic liquid crystals; thermotropic; smectic phase A; differential scanning calorimetry; polarizing optical microscopy; thermogravimetric analysis

1. Introduction

The 1,1′-dialkyl-4,4′-bipyridium salts are commonly known as viologens. They are an important class of dicationic salts and appropriately called advanced functional materials. The versatility of their applications arises from their redox properties, ionic conductivity, thermochromism, photochromism, and electrochromism [1]. These applications, to name a few, include electrochromic devices, molecular machines, organic batteries, and carbohydrate oxidation catalysts in alkaline fuel cells [2,3]. In addition, they were studied not only for the preparation of the ever-increasing class of ionic liquids with proper chemical modifications of cations and anions [4–12], but also for the preparation of ionic liquid crystals (ILCs) [7,13–24]. To mention several their chemical structures (**I-VI**) are given in Figure 1 as representative examples. Their LC phases are found to be dependent on both the chemical architecture of viologen moieties including dialkyl [13,17–19,23], asymmetric dialkyl [7], di(oligooxyethylene) [14,15], di(3,4,5-tri-n-alkoxybenbenzyl) [22], diphenyl [16,24], and 4-n-alkoxyldiphenyl [24], and the chemical architecture of the anions are of varying sizes. The anions to date include bromide, iodide [14,15], $^-BF_4$, $^-PF_6$ [22,24], ^-OTf, ^-SCN, $^-NTf_2$, 4-n-alkylsulfonates [24], 4-n-alkylbenzenesulfonates [16], and 3,4,5-tri-n-dodecyloxybenzenesulfonate [23]. Recently, their use in energy-related systems stemming from the unique properties of redox properties as well as LC properties of viologen moieties has given the boost for the exploration of this class of materials. Undoubtedly, this field is an active area of research for a decade or so that is manifested in a number of excellent reviews on this important topic [2,25–31].

As a continuation of our research efforts in the ILCs, herein, we describe the synthesis of a series of symmetric viologen compounds with 4-n-alkylbenzenesulfonates (n = 6, 7, 8, 9, 10, 12, 14, 16, 18);

wherein n denotes the carbon atoms in the alkyl chain), determine both their chemical structures by ^{1}H and ^{13}C NMR spectra—as well as elemental analysis—and the characterization of their thermotropic LC properties by several experimental techniques, including differential scanning calorimetry (DSC) and polarizing optical microscopy (POM). Their thermal stabilities by thermogravimetric analysis (TGA) are also included. The general structures and designations for these synthesized viologen salts, and their synthetic routes are shown in Scheme 1. The LC properties of this series of symmetric viologen salts with these anions enable one to establish the structure-property relationship of this important class of ILCs. Additionally, in contrast to other anions, reports of ILCs on these anions are relatively less studied [16].

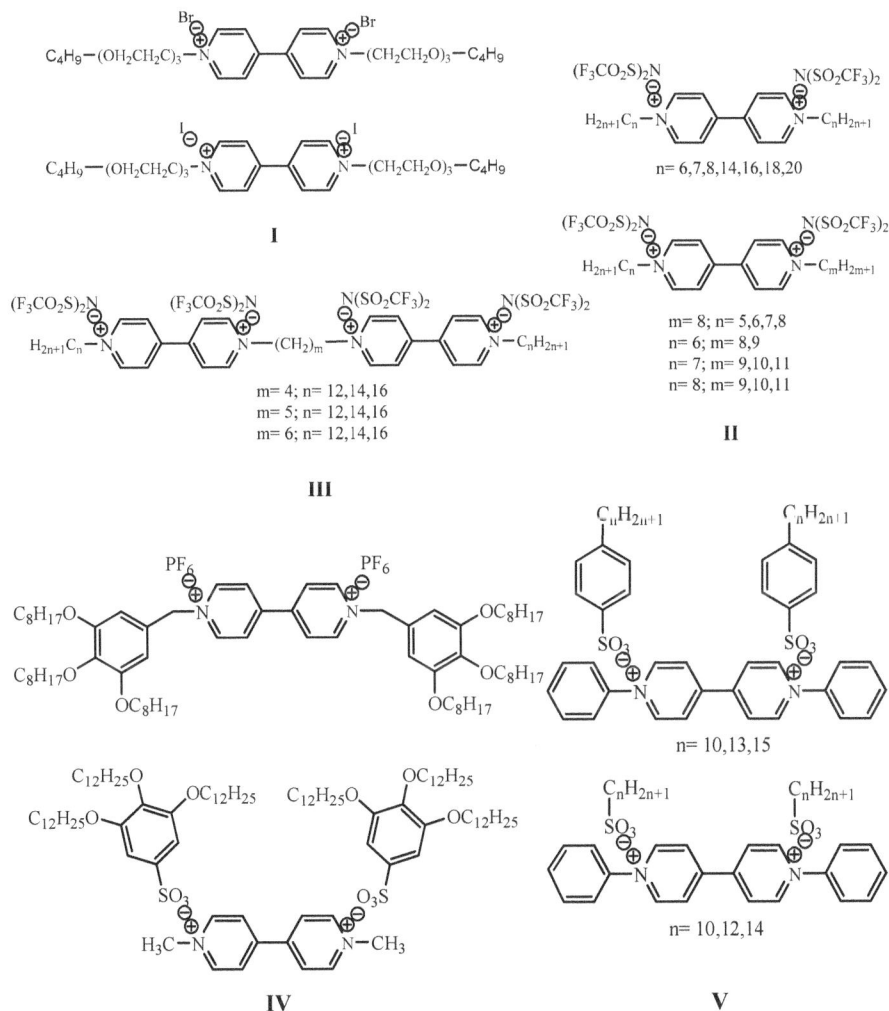

Figure 1. *Cont.*

13

n= 10,12,14

n= 10,12,14

$$Y= PF_6, BF_4, SO_3CF_3, SCN$$

VI

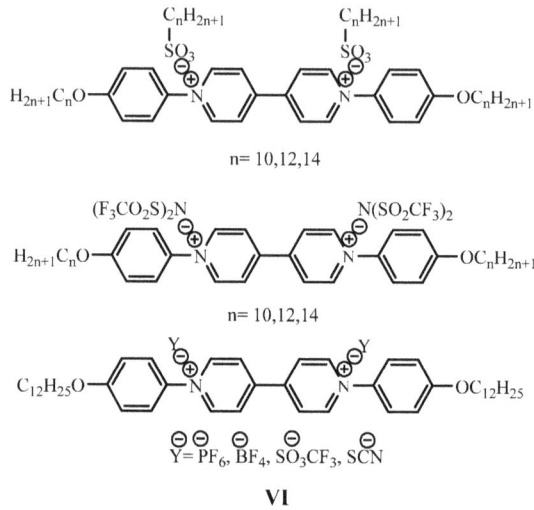

Figure 1. Some of the representative ILCs based on viologens that exhibits smectic phases and columnar LC phases depending on their chemical architectures.

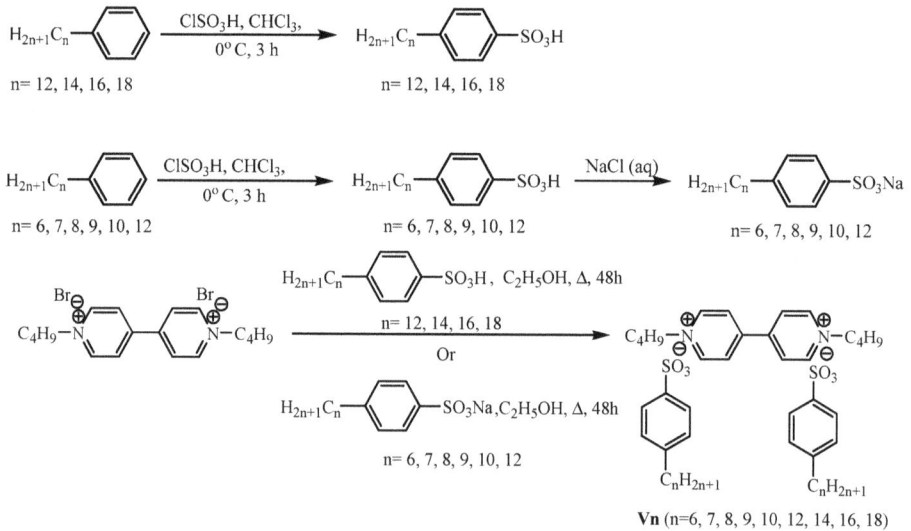

Scheme 1. Synthetic routes for the preparation of symmetric viologen salts containing 4-n-alkylbenzenesulfonates (Vn).

2. Materials and Methods

2.1. Instrumentation

The Fourier transform infrared (FTIR) spectra of several viologen salts were recorded with a Shimadzu IRPrestige FTIR analyser with their neat films on KBr pellets. The ^{1}H and ^{13}C nuclear magnetic resonance (NMR) spectra of the symmetric viologen salts, whenever possible, in CD$_3$OD were recorded by using VNMR 400 spectrometer operating at 400 and 100 MHz at room temperature. Elemental analysis was performed by Atlanta Microlab Inc., Norcross, GA. Differential scanning calorimetry (DSC) measurements of these salts were conducted on TA module DSC Q200 series in nitrogen at heating and cooling rates of 10 °C/min. The temperature axis of the DSC thermograms was calibrated before use with reference standards of high purity indium and tin. Their thermogravimetric analyses (TGA) were performed using a TGA Q50 instrument at a heating rate of 10 °C/min in nitrogen. Polarizing optical microscopy (POM) studies of these salts were performed by sandwiching each of them between a standard microscope glass slide and coverslip. The salts were heated and cooled on a Mettler hotstage (FP82HT) and (FP90) controller; and observations of the phases were made between crossed polarizers of an Olympus BX51 microscope. In short, salts were heated above their clearing transitions and cooled at 10 °C/min to room temperature, with brief pauses to collect images and observe specific transitions.

2.2. General Procedure for the Synthesis of 4-n-alkylbenzenesulfonic Acids (n = 12, 14, 16, 18)

The synthesis of 4-n-alkylbenzenesulfonic acids was carried out in accordance with the literature procedures [32,33]. The preparation of dodecylbenzenesulfonic acid from n-dodecylbenzene by using chlorosulfonic acid in chloroform was described as a typical procedure [33].

Chloroform (20 mL) was added to n-dodecylbenzene (2.26 g, 9.17 mmol) to form a colorless solution in a 50 mL flask that was placed in an ice bath. Cholorosulfonic acid (1.28 g, 11.0 mmol) was added slowly to the reaction flask over 10 min, resulting in a yellow viscous solution. The reaction mixture was stirred for 3 h inside an ice bath. At the end of the reaction, the reaction flask was taken out of the ice bath to reach the ambient temperature. The organic solvent was removed using a rotary evaporator. The product was dried overnight in vacuum to remove any residual solvent. It was then purified by washing with the excess toluene and dried again in vacuum to yield the white product (2.20 g, 73%). Similarly, 4-n-tetradecylbenzene- (2.96 g, 8.35 mmol), 4-n-hexadecylbenzene (2.69 g, 7.04 mmol), and 4-n-octadecylbenzenesulfonic (2.79 g, 6.80 mmol) were prepared from the corresponding n-alkylbenzenes with the yields of 74, 67, and 70%, respectively.

2.3. General Procedure for the Synthesis of Sodium 4-n-alkylbenzenesulfonates (n = 6, 7, 8, 9, 10, 12)

The modified literature procedure [34] was adopted for the synthesis of sodium 4-n-octylbenzenesulfonate, as an example, is as follows. Chloroform (20 mL) was added to n-octylbenzene (2.60 g, 13.7 mmol) to form a colorless solution in an Erlenmeyer flask that was set in an ice bath. Chlorosulfonic acid (1.90 g, 16.3 mmol) was added dropwise to the reaction flask, forming a yellow solution. The reaction mixture was stirred for 3 h in an ice bath. At the end of the reaction, the reaction flask was taken out of the ice bath to reach the ambient temperature. The organic solvent from the reaction mixture was removed using a rotary evaporator to yield the yellow viscous liquid. A saturated sodium chloride solution was carefully added to the viscous liquid on stirring to form the white precipitate of the product. It was then filtered and washed with a minimum quantity of water and excess toluene, respectively, to yield the white product (2.91 g, 9.95 mmol, 73%). Similarly, sodium 4-n-hexylbenzenesulfonate (3.02 g, 11.4 mmol), sodium 4-n-heptylbenzenesulfonate (2.91 g, 10.5 mmol), sodium 4-n-nonylbenzenesulfonate (3.00 g, 9.79 mmol), sodium 4-n-decylbenzenesulfonate (2.30 g, 7.18 mmol), and sodium 4-n-dodecylbenzenesulfonate (2.90 g, 8.85 mmol) were prepared from the corresponding n-alkylbenzenes with the yields of 75, 73 74, 57, and 76%, respectively.

2.4. Synthesis of 1,1′-di-n-butyl-4,4′-bipridinium Dibromide

This salt was prepared according to the literature procedure [18].

2.5. General Procedure for the Synthesis1,1′-di-n-butyl-4,4′-bipyridinium di(4-n-alkylbenzenesulfonates) by Metathesis Reaction (V12, V14, V16, V18) Using 4-n-alkylbenzenesulfonic Acid [35]

Ethanol (35 mL) was slowly added to 1,1′-di-n-butyl-4,4′-bipyridinium dibromide (0.93 g, 2.16 mmol) in a 100 mL round-bottomed flask and heated on stirring to form a clear solution. The 4-dodecylbenzenesulfonic acid (1.50 g, 4.60 mmol) was then slowly added to the hot solution, and the reaction flask was heated to reflux for 48 h. The solvent was reduced using a rotary evaporator. Then cold water was added to dissolve the HBr acid produced in the metathesis reaction to yield the product as white precipitate. It was collected by filtration and washed with excess water until the filtrate was neutral to litmus paper to yield the white product. Finally, **V12** was washed with ether and dried in vacuum to yield (1.77 g, 1.88 mmol, 87%). IR (KBr) ν (cm^{-1}): 3507, 3433, 3125, 3051, 2959, 2920, 2851, 1640, 1601, 1562, 1466, 1450, 1404, 1377, 1238, 1211, 1188, 1123, 1034, 1011, 953, 891, 841, 822, 779, 718, 694, 675, 610. ^{1}H NMR (CD$_3$OD, 400 MHz, ppm) δ 9.24 (d, *J* = 6.8 Hz, 4H), 8.66 (d, *J* = 6.8 Hz, 4H), 7.71 (d, *J* = 8.4 Hz, 4H), 7.23 (d, *J* = 8.4 Hz, 4H), 4.74 (t, *J* = 7.6 Hz, 4H), 2.63 (t, J= 7.6 Hz, 4H), 2.08–2.01 (m, 4H), 1.63–1.59 (m, 4H), 1.48–1.41 (m, 4H), 1.32–1.28 (m, 36H), 1.02 (t, *J* = 7.4 Hz, 6H), 0.89 (t, *J* = 7.4 Hz, 6H). ^{13}C NMR (CD$_3$OD, 100 MHz, ppm): δ 149.79, 145.62, 145.23, 142.42, 127.89, 126.85, 125.52, 61.61, 35.21, 33.00, 31.65, 31.14, 29.36, 29.33, 29.30, 29.16, 29.05, 28.87, 22.31, 19.03, 13.04, 12.34. Anal. Calcd. for C$_{54}$H$_{84}$N$_2$O$_6$S$_2$·H$_2$O (939.40): C 69.04, H 9.23, N 2.98, S 6.83; found: C 69.03, H 9.41, N 2.97, S 6.76. Similarly, **V14**, **V16**, and **V18** were prepared by using the corresponding 4-n-alkylbenzenesulfonic acid with the yields of 93, 92, and 93%, respectively. Data for **V14**: ^{1}H NMR (CD$_3$OD, 400 MHz, ppm) δ 9.24 (d, *J* = 6.8 Hz, 4H), 8.66 (d, *J* = 6.8 Hz, 4H), 7.71 (d, *J* = 8.4 Hz, 4H), 7.23 (d, *J* = 8.4 Hz, 4H), 4.74 (t, *J* = 7.6 Hz, 4H), 2.63 (t, J= 7.6 Hz, 4H), 2.07–2.02 (m, 4H), 1.63–1.59 (m, 4H), 1.49–1.43 (m, 4H), 1.32–1.28 (m, 44H), 1.02 (t, *J* = 7.4 Hz, 6H), 0.89 (t, *J* = 7.4 Hz, 6H). ^{13}C NMR (CD$_3$OD, 100 MHz, ppm): δ 149.79, 145.63, 145.23, 142.42, 127.87, 126.85, 125.51, 61.63, 35.20, 33.00, 31.64, 31.13, 29.36, 29.33, 29.15, 29.04, 28.86, 22.30, 19.03, 13.02, 12.36 (Figure S7 in the Supplementary Materials). Anal. Calcd. for C$_{58}$H$_{94}$N$_2$O$_6$S$_2$·H$_2$O (995.51): C 69.98, H 9.52, N 2.81, S 6.44; found: C 70.35, H 9.58, N 2.83, S 6.39. Data for **V16**: ^{1}H NMR (CD$_3$OD, 400 MHz, ppm) δ 9.24 (d, *J* = 6.8 Hz, 4H), 8.66 (d, *J* = 6.8 Hz, 4H), 7.71 (d, *J* = 8.4 Hz, 4H), 7.23 (d, *J* = 8.4 Hz, 4H), 4.74 (t, *J* = 7.6 Hz, 4H), 2.63 (t, *J* = 7.6 Hz, 4H), 2.08–2.01 (m, 4H), 1.63–1.59 (m, 4H), 1.48–1.41 (m, 4H), 1.32–1.28 (m, 52H), 1.02 (t, *J* = 7.4 Hz, 6H), 0.89 (t, *J* = 7.4 Hz, 6H). Because of limited solubility in CD$_3$OD, its ^{13}C NMR was not recorded (Figure S8). Anal. Calcd. for C$_{62}$H$_{100}$N$_2$O$_6$S$_2$·H$_2$O (1051.61): C 70.81, H 9.78, N 2.66, S 6.10; found: C 70.57, H 9.81, N 2.53, S 6.04. Data for **V18**: ^{1}H NMR (CD$_3$OD, 400 MHz, ppm) δ 9.24 (d, *J* = 6.8 Hz, 4H), 8.66 (d, *J* = 6.8 Hz, 4H), 7.71 (d, *J* = 8.4 Hz, 4H), 7.23 (d, *J* = 8.4 Hz, 4H), 4.74 (t, *J* = 7.6 Hz, 4H), 2.64 (t, *J* = 7.6 Hz, 4H), 2.07–2.04 (m, 4H), 1.63–1.59 (m, 4H), 1.49–1.44 (m, 4H), 1.32–1.28 (m, 60H), 1.03 (t, *J* = 7.4 Hz, 6H), 0.89 (t, *J* = 7.4 Hz, 6H). Because of limited solubility in CD$_3$OD, its ^{13}C NMR was not recorded (Figure S9). Anal. Calcd. for C$_{66}$H$_{110}$N$_2$O$_6$S$_2$·H$_2$O (1107.72): C 71.56, H 10.01, N 2.53, S 5.79; found: C 71.35, H 10.14, N 2.51, S 5.72.

2.6. General Procedure for the Synthesis 1,1′-di-n-butyl-4,4′-bipyridinium di(4-n-alkylbenzenesulfonates) by Metathesis Reaction (V6, V7, V8, V9, V10, V12,) Using Sodium 4-n-alkylbenzenesulfonate

Ethanol (35 mL) was slowly added to 1,1′-di-n-butyl-4,4′-bipyridinium dibromide (0.47 g, 1.09 mmol) in a 100 mL round-bottomed flask and heated on stirring to form a clear solution. Sodium 4-n-dodecylbenzenesulfonate (0.83 g, 2.40 mmol) was then slowly added to the hot solution, and the reaction flask was then heated to reflux for 48 h. The solvent was reduced using a rotary evaporator. Then cold water was added to dissolve the NaBr produced in the metathesis reaction to yield the product as a white precipitate. It was collected by filtration and washed with excess water until the filtrate gave the negative test with aqueous AgNO$_3$ solution to yield the white product. Finally, **V12** was washed with ether and dried in vacuum to yield (0.68 g, 0.74 mmol, 68%). Its spectral characteristics and elemental analysis were identical to those of **V12** prepared by using 4-n-dodecylbenzenesulfonic acid.

Similarly, in the cases of **V6–V10**, methanol (rather than ethanol) was used for this metathesis reaction for better solubility, by using the corresponding sodium 4-n-alkylbenzenesulfonates. Their yields were (1.77 g, 88%), (1.89 g, 93%), (1.25 g, 89%), (1.90 g, 96%), and (1.50 g, 75%), respectively. Data for **V6**: ^{1}H NMR (CD$_3$OD, 400 MHz, ppm) δ 9.24 (d, *J* = 6.8 Hz, 4H), 8.66 (d, *J* = 6.8 Hz, 4H), 7.71 (d, *J* = 8.4 Hz, 4H), 7.24 (d, *J* = 8.4 Hz, 4H), 4.74 (t, *J* = 7.6 Hz, 4H), 2.65 (t, J= 7.6 Hz, 4H), 2.09–2.01 (m, 4H), 1.63–1.57 (m, 4H), 1.50–1.41 (m, 4H), 1.36–1.30 (m, 12H), 1.04 (t, *J* = 7.4 Hz, 6H), 0.89 (t, *J* = 7.4 Hz, 6H). ^{13}C NMR (CD$_3$OD, 100 MHz, ppm): δ 149.79, 145.62, 145.23, 142.39, 127.88, 126.84, 125.49, 61.62, 35.19, 32.99, 31.38, 31.07, 28.50, 22.22, 19.02, 12.96, 12.37, 12.36. Anal. Calcd. for C$_{42}$H$_{60}$N$_2$O$_6$S$_2$·H$_2$O

(771.08): C 65.42, H 8.10, N 3.63, S 8.32; found: C 65.29, H 8.12, N 3.55, S 8.27. Data for **V7**: ^{1}H NMR (CD$_3$OD, 400 MHz, ppm) δ 9.24 (d, J = 6.8 Hz, 4H), 8.66 (d, J = 6.8 Hz, 4H), 7.71 (d, J = 8.4 Hz, 4H), 7.24 (d, J = 8.4 Hz, 4H), 4.74 (t, J = 7.6 Hz, 4H), 2.65 (t, J= 7.6 Hz, 4H), 2.06–2.03 (m, 4H), 1.63–1.59 (m, 4H), 1.48–1.43 (m, 4H), 1.33–1.25 (m, 16H), 1.04 (t, J = 7.4 Hz, 6H), 0.89 (t, J = 7.4 Hz, 6H). ^{13}C NMR (CD$_3$OD, 100 MHz, ppm): δ 149.79, 145.62, 145.23, 142.39, 127.88, 126.84, 125.49, 61.62, 35.18, 32.99, 31.53, 31.11, 28.81, 22.24, 19.02, 12.98, 12.36. Anal. Calcd. for C$_{44}$H$_{64}$N$_2$O$_6$S$_2$·H$_2$O (799.13): C 66.13, H 8.32, N 3.51, S 8.02; found: C 66.20, H 8.20, N 3.52, S 7.93. Data for **V8**: IR (KBr) ν (cm^{-1}): 3507, 3433, 3125, 3051, 2955, 2924, 2851, 1643, 1562, 1450, 1377, 1238, 1211, 1188, 1119, 1034, 1011, 845, 690, 606. ^{1}H NMR (CD$_3$OD, 400 MHz, ppm) δ 9.24 (d, J = 6.8 Hz, 4H), 8.66 (d, J = 6.8 Hz, 4H), 7.69 (d, J = 8.4 Hz, 4H), 7.22 (d, J = 8.4 Hz, 4H), 2.65 (t, J= 7.6 Hz, 4H), 2.09-2.01 (m, 4H), 1.63–1.59 (m, 4H), 1.51–1.41 (m, 4H), 1.30–1.26 (m, 20H), 1.00 (t, J = 7.4 Hz, 6H), 0.87 (t, J = 7.4 Hz, 6H). ^{13}C NMR (CD$_3$OD, 100 MHz, ppm): δ 149.79, 145.62, 145.24, 142.41, 127.89, 126.85, 125.51, 61.63, 35.20, 33.00, 31.58, 31.12, 29.11, 28.97, 28.85, 22.28, 19.04, 19.03, 13.01, 12.38. Anal. Calcd. for C$_{46}$H$_{68}$N$_2$O$_6$S$_2$·H$_2$O (827.19): C 66.79, H 8.53, N 3.39, S 7.75; found: C 66.56, H 8.73, N 3.39, S 7.76. Data for **V9**: ^{1}H NMR (CD$_3$OD, 400 MHz, ppm) δ 9.24 (d, J = 6.8 Hz, 4H), 8.66 (d, J = 6.8 Hz, 4H), 7.71 (d, J = 8.4 Hz, 4H), 7.24 (d, J = 8.4 Hz, 4H), 4.74 (t, J = 7.6 Hz, 4H), 2.65 (t, J = 7.6 Hz, 4H), 2.09–2.01 (m, 4H), 1.63–1.59 (m, 4H), 1.49–1.43 (m, 4H), 1.32–1.28 (m, 24H), 1.03 (t, J = 7.4 Hz, 6H), 0.89 (t, J = 7.4 Hz, 6H). ^{13}C NMR (CD$_3$OD, 100 MHz, ppm): δ 149.81, 145.64, 145.23, 142.38, 127.88, 126.85, 125.51, 61.64, 35.20, 33.01, 31.61, 31.12, 29.25, 29.15, 28.99, 28.84, 22.29, 19.03, 13.00, 12.37. Anal. Calcd. for C$_{48}$H$_{72}$N$_2$O$_6$S$_2$·2H$_2$O (873.26): C 66.02, H 8.77, N 3.21, S 7.34; found: C 66.19, H 8.62, N 2.90, S 7.94. Data for **V10**: IR (KBr) ν (cm^{-1}): 3507, 3433, 3125, 3051, 2959, 2920, 2851, 1639, 1450, 1238, 1211, 1188, 1122, 1034, 1011, 845, 694, 610. ^{1}H NMR (CD$_3$OD, 400 MHz, ppm) δ 9.24 (d, J = 6.8 Hz, 4H), 8.65 (d, J = 6.8 Hz, 4H), 7.71 (d, J = 8.4 Hz, 4H), 7.23 (d, J = 8.4 Hz, 4H), 4.74 (t, J = 7.6 Hz, 4H), 2.63 (t, J= 7.6 Hz, 4H), 2.08–2.00 (m, 4H), 1.62–1.59 (m, 4H), 1.50–1.41 (m, 4H), 1.31–1.28 (m, 28H), 1.03 (t, J = 7.4 Hz, 6H), 0.89 (t, J = 7.4 Hz, 6H). ^{13}C NMR (CD$_3$OD, 100 MHz, ppm): δ 149.84, 145.61, 145.23, 142.35, 127.86, 126.84, 125.49, 61.63, 35.18, 33.01, 31.61, 31.10, 29.27, 29.13, 29.00, 28.82, 22.28, 19.02, 12.99, 12.34. Anal. Calcd. for C$_{50}$H$_{76}$N$_2$O$_6$S$_2$·H$_2$O (883.29): C 67.99, H 8.90, N 3.17, S 7.26; found: C 67.73, H 8.90, N 3.17, S 7.26 (Figure S1–S6).

3. Results and Discussion

In this study, several viologen salts containing 4-n-alkylbenzenesulfonates (**V6–V18**) were synthesized, characterized for their chemical structures by spectroscopic methods, and further characterized for their thermotropic LC properties by DSC and POM studies. The thermal stabilities of the viologen salts were also determined by TGA.

3.1. Synthesis of Viologen Salts (V6–V18)

The synthetic methods of 4-n-alkylbenzenesulfonic acids (n = 12, 14, 16, 18) and sodium 4-n-alkylbenzenesulfonates (n = 6, 7, 8, 9, 10, 12) are shown in Scheme 1, which also includes the synthetic procedures of the viologen salts (**V6–V18**). We prepared the long alkyl chain sulfonic acids as hydrates in respectable yields, in contrast, we also prepared the short alkyl chain sodium 4-n-alkylbenzenesulfonates as anhydrous forms in respectable yields. Both sulfonic acids and sodium salts were successfully used in the metathesis reaction for the synthesis of **V6–V18** salts as monohydrates except **V9**, in this case, it was dihydrate from viologen dibromides. The yields of the metathesis reaction were also respectable.

3.2. Thermotropic LC Properties of 4-n-alkylbenzenesulfonic Acids (n = 12, 14, 16, 18) by DSC and POM [36–40]

Sodium 4-n-alkylbenzenesulfonates (n = 6, 7, 8, 9, 10) showed melting transitions, that is, solid to isotropic transitions with high melting enthalpies at relatively high temperatures, as expected. Their melting peaks were at 296, 285, 275, 263, and 261 °C, respectively, as determined by DSC at a heating rate of 10 /min. Their thermal stability was in the range of 413–429 °C as determined by TGA at a heating rate of 10 °C/min in nitrogen. In contrast, 4-n-alkylbenzenesulfonic acids (n = 12, 14, 16, 18), as hydrates showed LC phases at relatively low melting temperatures and also showed isotropic transitions at high temperatures. Their melting peaks were at 40, 47, 52, and 65 °C, respectively, as determined by DSC at a heating rate of 10 °C/min, and isotropic peaks were at 134, 150, 144, and 141 °C, respectively. Figure 2 shows the LC phases of n = 12 and n = 18 sulfonic acids as examples. Their thermal stability was relatively low: as expected, there was 15% of water at about 130 °C.

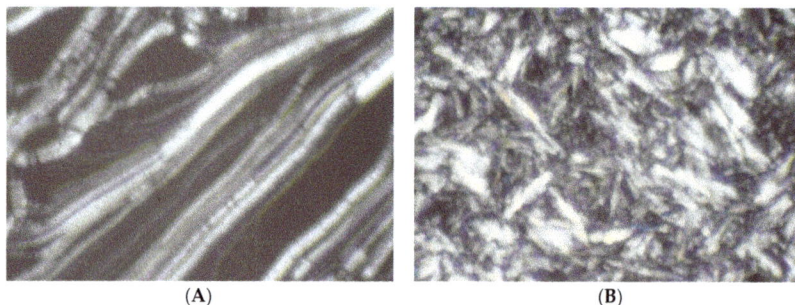

| (A) | (B) |

Figure 2. POM textures of **(A)** 4-n-dodecylbenzenesulfonic acid at 100 °C and **(B)** 4-n-octadecylbenzenesulfonic acid at 110 °C displaying oily streaks and bâtonnets textures of SmA phases (magnification 400 ×).

3.3. Thermotropic LC Properties of (V6-V18) by DSC and POM [36–40]

Figure 3 displays the DSC thermograms of **V6** in the heating and cooling cycles. The first heating cycle clearly shows two endotherms. In conjunction with POM, the large endotherm corresponded to crystal–LC transition, T_m, at 185 °C, and the small endotherm corresponded to LC–isotropic transition, T_i, at 202 °C. In the first cooling cycle, it showed an exotherm that corresponded to the isotropic transition to LC transition. The absence of LC-crystallization exotherm in the first cooling cycle and the crystal-LC transition in the second heating cycle suggested it remained in the LC state and subsequently went to isotropization at 213 °C. Thus, it was found that **V6** showed the LC phase that underwent isotropic transition at high temperature. The DSC thermograms of **V7** were essentially identical to those of **V6** and hence the similar interpretations, suggesting that it also showed a T_m at 185 and a T_i at 199 °C. It went to isotropization at 214 °C in the second heating cycle. Like **V6**, **V7** formed a LC phase and an isotropic phase (Figure S10).

Figure 4 shows the DSC thermograms of **V8** in its heating and cooling cycles. In the first heating cycle, it showed three endotherms that were related to the crystal–LC phase (T_m) at 181, LC–LC transition at 199, and LC-isotropic transition at 238 °C. In the first cooling cycle, isotropic-SmA, SmA-SmX, and Smx-crystalline transitions occurred at 227, 194, and 127 °C, respectively. Figure 5 shows the DSC thermograms **V9** in its heating and cooling cycles. In the first heating cycle, it showed the T_m at 178 and T_i at 268 °C. In between T_m and T_i there were two additional endotherms that were presumably related LC–LC transitions. In the first cooling cycle, the isotropic–SmA, SmA–crystal and crystal–crystal transition occurred at 250, 161, and 39 °C, respectively. The features for DSC thermograms of **V10–V18** in their heating and cooling cycles were essentially identical (Figure S11–S14). For example, **V10** showed T_m at 174 wherein it transformed into a LC phase and T_i at 202 °C as a small endotherm in the first heating cycle. In the first cooling cycle, the isotropic–LC transition was not detected in its DSC thermogram, but it was detected by POM studies. However, the LC–crystal transition was detected in the first cooling cycle and verified with POM studies. It did not show the crystal–crystal transition at low temperature, but **V12–V18** did show addition endotherms as these transitions at low temperatures. Figure 6 shows the DSC thermograms of **V18** in its heating and cooling cycles, which are the representative thermograms for **V10–V16**. Figure 7 shows the photomicrographs of **V8, V9, V10,** and **V12** taken at specified temperatures suggestive of their SmA phases (Figure S15).

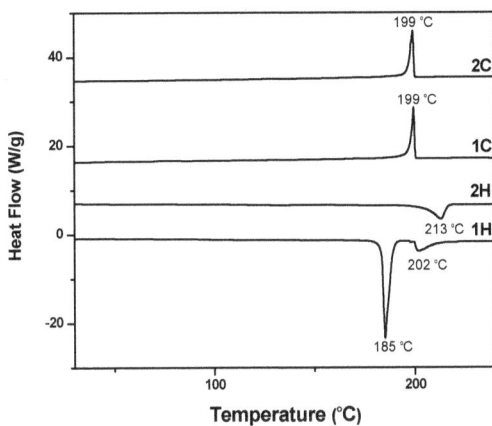

Figure 3. DSC thermograms of **V6** obtained at heating and cooling rates of 10 °C/min in nitrogen.

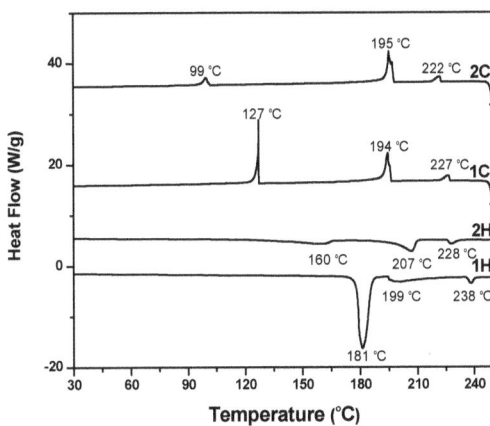

Figure 4. DSC thermograms of **V8** obtained at heating and cooling rates of 10 °C/min in nitrogen.

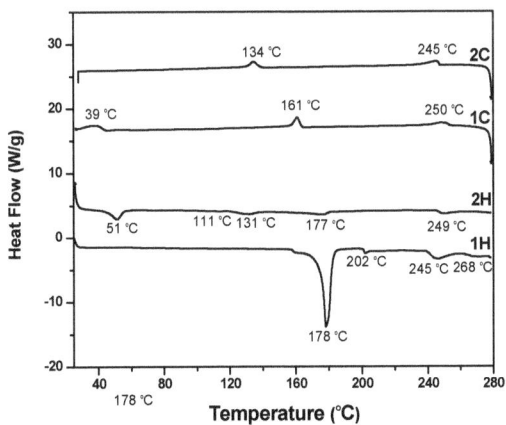

Figure 5. DSC thermograms of **V9** obtained at heating and cooling rates of 10 °C/min in nitrogen.

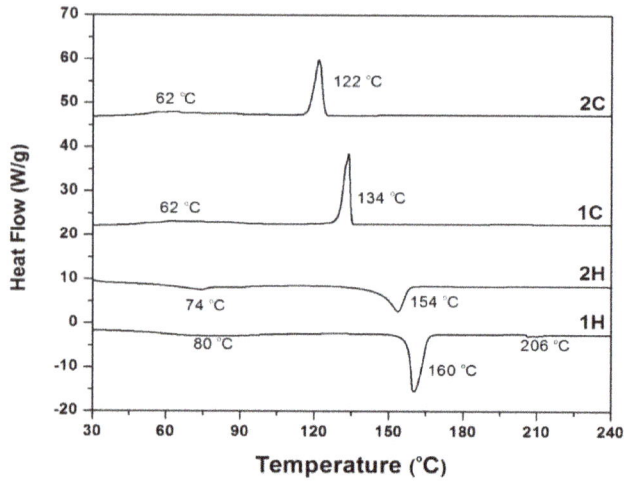

Figure 6. DSC thermograms of **V18** obtained at heating and cooling rates of 10 °C/min in nitrogen.

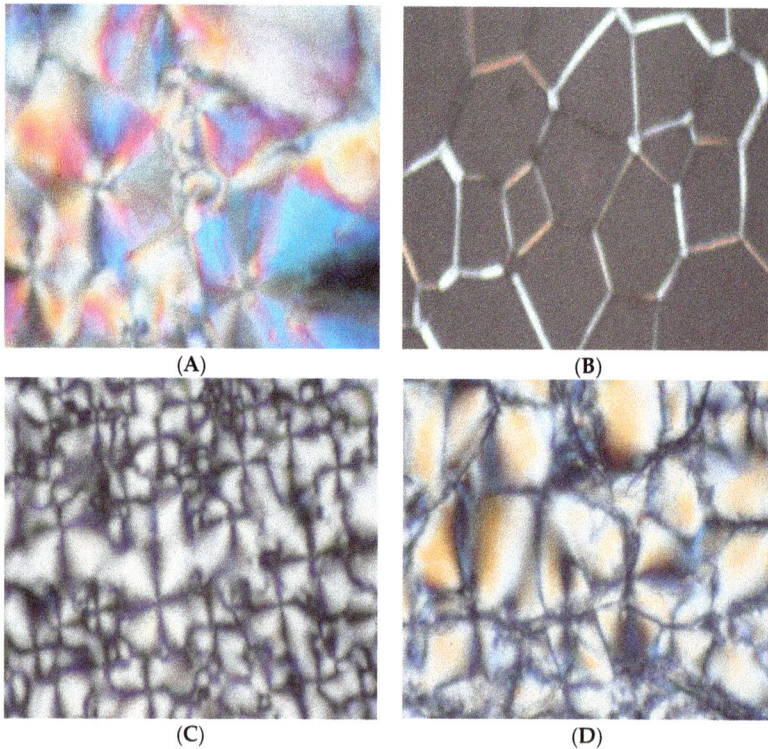

Figure 7. Typical textures observed by polarizing optical microscopy studies, revealing Schlieren or focal conic textures of (**A**) **V8** at 230 °C (**B**) **V9** at 260 °C (**C**) **V10** at 190 °C (**D**) **V12** at 200 °C on heating suggestive of their smectic A LC phases (magnification 400 ×).

The thermodynamic properties of phase transition temperatures of **V6–V18** determined from DSC measurements and POM studies are compiled in Table 1.

Table 1. Thermodynamic properties of phase transition temperatures of **V6–V18** obtained from DSC measurements and POM textures. Phase transition temperatures (°C) and their enthalpy changes (J/g) were taken at a scanning rate of 10 °C/min from the first heating and cooling cycles.

Identification	Phase Transition Temperature (Enthalpy Change) °C (J/g)
V6	Cr 185 (85.4) SmA 202 (14.3) I
	I 199 (−26.9) SmA
V7	Cr 185 (84.0) SmA 199 (24.1) I
	I 204 (−27.7) SmA
V8	Cr 181 (83.0) SmX 199 (12.7) SmA 238 (3.2) I
	I 227 (−3.3) SmA 194 (−17.9) SmX 127 (−12.4) Cr
V9	Cr 178 (85.8) SmX 202 (1.2) SmX 245 (10.6) SmA 268 (1.7) I
	I 250 (−5.0) SmA 161 (−7.0) Cr 39 (−7.6) Cr
V10	Cr 174 (84.2) SmA 202 (1.2) I
	I − SmA 148 (−71.8) Cr
V12	Cr 56 * Cr 169 (79.5) 206 (0.8) I
	I − SmA 145 (−58.3) Cr 54 Cr
V14	Cr 68 * Cr 164 (81.5) SmA 195 (20.9) I
	I − SmA 159 (−3.4) SmX 126 (−37.2) Cr 88 (−1.0) Cr 66 (−8.2) Cr
V16	Cr 75 * Cr 157 (81.3) SmA 205 (2.3) I
	I − SmA 160 (− 4. 2) SmX 113 (−29.4) Cr 73 (−20.0) Cr
V18	Cr 80 * Cr 160 (82.9) SmA 206 (2.2) I
	I − SmA 134 (−60.5) Cr 62 (−20.2) Cr

Cr-Crystal, I-Isotropic, SmA-Smectic A, SmX-Unidentified smectic phase. * Too broad transition.

The thermotropic LC properties of this new series of viologen salts are remarkable in the sense that they showed isotropic transitions below their decomposition temperatures. These results were presumably related to the flexible n-butyl groups lined to the viologen moieties. Note here that diphenylviologens containing the long alkylbenzene sulfonates (n = 10, 13 and 15) do form SmA phases until their decomposition temperatures, that is, they do not show the isotropic transitions because of the presence of the more rigid phenyl groups attached to the viologen moieties [16,35]. The short alkylbenzenesulfonates with n = 6 and 8 do not form LC phases [16,35]. In addition, guanidinium alkylbezene sulfonates with even alkyl chain lengths n = 8 or higher do form smectic LC phases [34]. This study along with other reports (vide supra) indicate that n-alkylbenzenesulfonates are an interesting class of counterions that may be exploited for the synthesis of ILCs including viologens-a versatile class of functional materials.

3.4. Thermal Stabilities of Viologen Salts (V6–V18)

The thermal stabilities of the viologen salts were studied by TGA and determined as the temperature (°C) at which a 5% weight loss for each of the salts occurred at a heating rate of 10 °C/min in nitrogen. Despite the presence of flexible alkyl chains both in the viologen moieties and the benzenesulfonate moieties, TGA thermograms of some of these salts (**V10–V18**), as shown in Figure 8, show relatively high thermal stabilities that are in the temperature range of 285–292 °C. On the one hand, these temperatures slightly increase at a gradual pace with the increase of carbon numbers in the alkyl chain. On the other hand, these temperatures (278–282 °C) for **V6–V8** gradually decrease with the decrease in carbon numbers in the alkyl chain, except for **V9** (295 °C) (Figure S16).

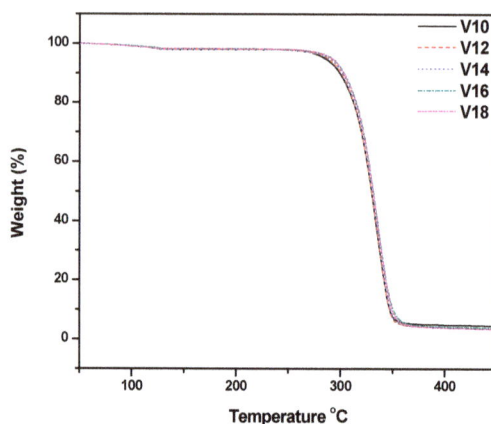

Figure 8. TGA thermograms of **V10–V18** obtained a heating rate of 10 °C/min in nitrogen.

4. Conclusions

A new series of viologens with 4-n-alkylbenzenesulfonates were prepared by the metathesis reaction of viologen dibromide with 4-n-alkylbenzenesulfonic acids or sodium-4-n-alkylbenzenesulfonates in alcohols. The chemical structures of these salts were established using spectroscopic techniques and elemental analysis. Their thermotropic LC properties as determined by DSC and POM suggested that they exhibited crystal–LC transitions and LC–isotropic transitions; and showed Schlieren or focal conic textures indicative of their SmA phases. They had good thermal stability. These results suggest that they belong to a class of ILCs that have practical implications.

Supplementary Materials: The following are available online at http://www.mdpi.com/2073-4352/9/2/77/s1, Figures S1–S7: ^{1}H and ^{13}C NMR spectra of V6–V14, Figures S8 and S9: ^{1}HNMR spectra of V16 and V18, Figures S10–S14: DSC thermograms of V7, V10, V12, V14 and V16, Figure S15: POM texture of V18, Figure S16: TGA thermograms of V6–V9.

Author Contributions: Conceptualization, P.K.B. and A.C.; Methodology, J.K., R.C.G.P.; Software, E.J.D.; Validation, H.H., A.C. and P.K.B.; Formal Analysis, H.H.; Investigation, H.H.; Resources, H.H.; Data Curation, E.J.D.; Writing-Original Draft Preparation, A.C.; Writing-Review & Editing, A.C.; Visualization, E.J.D.; Supervision, P.K.B.; Project Administration, P.K.B.

Funding: There is no external funding for this research.

Conflicts of Interest: The authors declare no conflict of interest.

References

1. Monk, P.M.S. *The Viologens Physiochemical Properties, Synthesis and Applications of the Salts of 4,4'-Bipyridine*; Wiley: New York, NY, USA, 1998; pp. 1–311.
2. Striepe, L.; Baumgartner, T. Viologens and Their Application as Functional Materials. *Chem. Eur. J.* **2017**, *23*, 16924–16940. [CrossRef] [PubMed]
3. Rigby, C.R.; Han, H.; Bhowmik, P.K.; Bahari, M.; Chang, A.; Harb, J.N.; Lewis, R.S.; Watt, G.D. Soluble Viologen Polymers as Carbohydrate Oxidation Catalysts for Alkaline Carbohydrate Fuel Cells. *J. Electroanal. Chem.* **2018**, *823*, 416–421. [CrossRef]
4. Hatazawa, T.; Terrill, R.H.; Murray, R.W. Microelectrode Voltammetry and Electron Transport in an Undiluted Room Temperature Melt of an Oligo(ethylene glycol)-Tailed Viologen. *Anal. Chem.* **1996**, *68*, 597–603. [CrossRef]
5. Ito-Akita, K.; Ohno, H. Low Temperature Molten Viologens-Phase Transitions and Electrochemical Properties. *Electrochem. Soc. Procced.* **1999**, *99-14*, 193–201.
6. Causin, V.; Saielli, G. Effect of a structural modification of the bipyridinium core on the phase behaviour of viologen-based bistriflimide salts. *J. Mol. Liq.* **2009**, *145*, 41–47. [CrossRef]

7. Causin, V.; Saielli, G. Effect of asymmetric substitution on the mesomorphic behaviour of low-melting viologen salts of bis(trifluoromethanesulfonyl)amide. *J. Mater. Chem.* **2009**, *19*, 9153–9162. [CrossRef]

8. Bonchio, M.; Carraro, M.; Casella, G.; Causin, V.; Rastrelli, F.; Saielli, G. Thermal behaviour and electrochemical properties of bis(trifluoromethanesulfonyl)amide and dodecatungstosilicate viologen dimers. *Phys. Chem. Chem. Phys.* **2012**, *14*, 2710–2717. [CrossRef]

9. Gunaratne, H.Q.N.; Nockemann, P.; Olejarz, S.; Reid, S.M.; Seddon, K.R.; Srinivasan, G. Ionic Liquids with Solvatochromic and Charge-Transfer Functionalities Incorporating the Viologen Moiety. *Aust. J. Chem.* **2013**, *66*, 607–611. [CrossRef]

10. Jordão, N.; Cabrita, L.; Pina, F.; Branco, L.C. Novel Bipyridinium Ionic Liquids as Liquid Electrochromic Devices. *Chem. Eur. J.* **2014**, *20*, 3982–3988. [CrossRef]

11. Tahara, H.; Furue, Y.; Suenaga, C.; Sagara, T. A Dialkyl Viologen Ionic Liquid: X-ray Crystal Structure Analysis of Bis(trifluoromethanesulfonyl)imide Salts. *Cryst. Growth Des.* **2015**, *15*, 4735–4740. [CrossRef]

12. Jordão, N.; Cruz, H.; Branco, A.; Pina, F.; Branco, L.C. Electrochromic Devices Based on Disubstituted Oxo-Bipyridinium Ionic Liquids. *ChemPlusChem.* **2015**, *80*, 202–208. [CrossRef]

13. Yu, L.P.; Samulski, E.T. Ionomeric Liquid Crystals. In *Oriented Fluids and Liquid Crystals*; Griffin, A.C., Johnson, J.F., Eds.; Plenum: New York, NY, USA, 1984; Volume 4, pp. 697–704.

14. Tabushi, I.; Yamamura, K.; Kominami, K. Electric Stimulus-Response Behavior of Liquid-Crystalline Viologen. *J. Am. Chem. Soc.* **1986**, *108*, 6409–6410. [CrossRef]

15. Yamamura, K.; Okada, Y.; Ono, S.; Kominami, K.; Tabushi, I. New Liquid Crystalline Viologens Exhibiting Electric Stimulus-Response Behavior. *Tetrahedron Lett.* **1987**, *28*, 6475–6478. [CrossRef]

16. Haramoto, Y.; Yin, M.; Matukawa, Y.; Ujiie, S.; Nanasawa, M. A new ionic liquid crystal compound with viologen group in the principal structure. *Liq. Cryst.* **1995**, *19*, 319–320. [CrossRef]

17. Bhowmik, P.K.; Han, H.; Cebe, J.J.; Burchett, R.A.; Acharya, A.; Kumar, S. Ambient temperature thermotropic liquid crystalline viologen bis(triflimide) salts. *Liq. Cryst.* **2003**, *30*, 1433–1440. [CrossRef]

18. Bhowmik, P.K.; Han, H.; Nedeltchev, I.K.; Cebe, J.J. Room-Temperature Thermotropic Ionic Liquid Crystals: Viologen Bis(triflimide) Salts. *Mol. Cryst. Liq. Cryst.* **2004**, *419*, 27–46. [CrossRef]

19. Bhowmik, P.K.; Killarney, S.T.; Li, J.R.A.; Koh, J.J.; Han, H.; Sharpnack, L.; Agra-Kooijman, D.M.; Fisch, M.R.; Kumar, S. Thermotropic liquid-crystalline properties of extended viologen bis(triflimide) salts. *Liq. Cryst.* **2018**, *45*, 872–885. [CrossRef]

20. Casella, G.; Causin, V.; Rastrelli, F.; Saielli, G. Viologen-based ionic liquid crystals: Induction of a smectic A phase by dimerisation. *Phys. Chem. Chem. Phys.* **2014**, *16*, 5048–5051. [CrossRef] [PubMed]

21. Casella, G.; Causin, V.; Rastrelli, F.; Saielli, G. Ionic liquid crystals based on viologen dimers: Tuning the mesomorphism by varying the conformational freedom of the ionic layer. *Liq. Cryst.* **2016**, *43*, 1161–1173. [CrossRef]

22. Tanabe, K.; Yasuda, T.; Yoshio, M.; Kato, T. Viologen-Based Redox-Active Ionic Liquid Crystals Forming Columnar Phases. *Org. Lett.* **2007**, *9*, 4271–4274. [CrossRef] [PubMed]

23. Asaftei, S.; Ciobanu, M.; Lepadatu, A.M.; Song, E.; Beginn, U. Thermotropic ionic liquid crystals by molecular assembly and ion pairing of 4,4′-bipyridinium derivatives and tris(dodecyloxy)benzenesulfonates in a non-polar solvent. *J. Mater. Chem.* **2012**, *22*, 14426–14437. [CrossRef]

24. Wang, R.-T.; Lee, G.-H.; Lai, C.K. Anion-induced ionic liquid crystals of diphenylviologens. *J. Mater. Sci. C* **2018**, *6*, 9430–9444. [CrossRef]

25. Binnemans, K. Ionic Liquid Crystals. *Chem. Rev.* **2005**, *105*, 4148–4204. [CrossRef] [PubMed]

26. Axenov, K.; Laschat, S. Thermotropic ionic liquid crystals. *Materials* **2011**, *4*, 206–259. [CrossRef] [PubMed]

27. Causin, V.; Saielli, G. Ionic liquid crystals. In *Green Solvents II. Properties and Applications of Ionic Liquids*; Mohammad, A., Inamuddin, D., Eds.; Springer: London, UK, 2012; pp. 79–118.

28. Mansueto, M.; Laschat, S. Ionic Liquid Crystals. In *Handbook of Liquid Crystals. Vol. 6: Nanostructured and Amphiphilic Liquid Crystals*, 2nd ed.; Goodby, J.W., Collings, P.J., Kato, T., Tschierske, C., Gleeson, H., Raynes, P., Eds.; Wiley-VCH: Weinheim, Germany, 2014; pp. 231–280.

29. Fernandez, A.A.; Kouwer, P.H.J. Key Developments in Ionic Liquid Crystals. *Int. J. Mol. Sci.* **2016**, *17*, 731. [CrossRef] [PubMed]

30. Goossens, K.; Lava, K.; Bielawski, C.W.; Binnemans, K. Ionic Liquid Crystals: Versatile Materials. *Chem. Rev.* **2016**, *116*, 4643–4807. [CrossRef] [PubMed]

31. Kato, T.; Yoshio, M.; Ichikawa, T.; Soberats, B.; Ohno, H.; Funahashi, M. Transport of ions and electrons in nanostructured liquid crystals. *Nat. Rev. Mater.* **2017**, *2*, 17001. [CrossRef]
32. Gilbert, E.E. Recent Developments in Preparative Sulfonation and Sulfation. *Synthesis* **1969**, *1969*, 3–10. [CrossRef]
33. Martin, S.M.; Yonezawa, J.; Horner, M.J.; Macosko, C.W.; Ward, M.D. Structure and rheology of hydrogen bond reinforced liquid crystals. *Chem. Mater.* **2004**, *16*, 3045–3055. [CrossRef]
34. Mathevet, F.; Masson, P.; Nicoud, J.-F.; Skoulios, A. Smectic liquid crystals from supramolecular guanidinium alkylbenzenesulfonates. *Chem. Eur. J.* **2002**, *8*, 2248–2254. [CrossRef]
35. Nanasawa, M.; Matsukawa, Y.; Jin, J.J.; Haramoto, Y. Redox photochemistry of viologen in organized solid state. *J. Photochem. Photobiol. A Chem.* **1997**, *109*, 35–38. [CrossRef]
36. Gray, G.W.; Goodby, J.W.G. *Smectic Liquid Crystals: Textures and Structures*; Leonard Hill: Glasgow, UK, 1984.
37. Collins, P.J.; Hird, M. *Introduction to Liquid Crystals Chemistry and Physics*; Taylor & Francis: Bristol, PA, USA, 1997.
38. Demus, D.; Goodby, J.W.; Gray, G.W.; Spiess, H.-W.; Vill, V. (Eds.) *Handbook of Liquid Crystals*; Wiley-VCH: Weinheim, Germany, 1998; Volumes 1–3.
39. Dierking, I. *Textures of Liquid Crystals*; Wiley-VCH: Weinheim, Germany, 2003.
40. Goodby, J.W.; Collings, P.J.; Kato, T.; Tschierske, C.; Gleeson, H.F.; Raynes, P. (Eds.) *Handbook of Liquid Crystals: 8 Volume Set*, 2nd ed.; Wiley-VCH: Weinheim, Germany, 2014.

crystals

MDPI

Article

Improved Electronic Transport in Ion Complexes of Crown Ether Based Columnar Liquid Crystals

Peter Staffeld [1], Martin Kaller [2], Philipp Ehni [2], Max Ebert [2], Sabine Laschat [2,*] and Frank Giesselmann [1,*]

[1] Institut für Physikalische Chemie, Universität Stuttgart, Pfaffenwaldring 55, D-70569 Stuttgart, Germany; peter.staffeld@googlemail.com

[2] Institut für Organische Chemie, Universität Stuttgart, Pfaffenwaldring 55, D-70569 Stuttgart, Germany; m.kaller@icloud.com (M.K.); philipp.ehni@oc.uni-stuttgart.de (P.E.); max.ebert@oc.uni-stuttgart.de (M.E.)

* Correspondence: sabine.laschat@oc.uni-stuttgart.de (S.L.); f.giesselmann@ipc.uni-stuttgart.de (F.G.); Tel.: +49-711-68564565 (S.L.); +49-711-68564460 (F.G.)

Received: 21 December 2018; Accepted: 29 January 2019; Published: 31 January 2019

Abstract: The Li^+- and K^+-complexes of new discotic mesogens, where two n-alkoxy-substituted triphenylene cores are connected by a central crown ether (12-crown-4 and 18-crown-6), provide interesting structural and electronic properties. The inter- and intra-columnar structure was investigated by small and wide angle X-ray scattering. The electronic and ionic transports were studied by temperature dependent photoconductivity and impedance spectroscopy, respectively. Besides a strong increase of the stability and the width of the columnar phases the presence of soft anions (iodide, thiocyanate, tetrafluoroborate) leads to an improved intra-columnar order. The hereby shortened stacking-distance of the triphenylene cores leads to a significant increase of the photoconductivity in the columnar mesophase. Furthermore, the ionic conductivity of the new materials was investigated on macroscopically aligned thin films. The existence of channels for fast cation transport formed by the stacked crown ether moieties in the centre of each column can be excluded. The cations are coordinated strongly and therefore contributing only little to the conductivity. The ionic conductivity is dominated by the anisotropic migration of the non-coordinated anions through the liquid, like side chains favouring the propagation parallel to the columns. Iodide migrates about 20 times faster than thiocyanate and 100 times faster than tetrafluoroborate.

Keywords: liquid crystals; columnar; discotic; crown ether; electron transport; ion transport; ion channels; impedance spectroscopy; photoconductivity; X-ray diffraction; salt effect

1. Introduction

Columnar phases of discotic liquid crystals display one-dimensional (1D) photoconductivity according to a seminal discovery by Haarer et al. in 1993 [1], turning the investigation of the electronic charge transport in columnar liquid crystals into a very active field of research [2,3]. Columnar materials can be used as semiconductors in applications, like organic field effect transistors (OFET) [4,5], organic light emitting diodes (OLED) [6], or organic photovoltaic cells (OPV) [7]. The main advantage of using organic materials with liquid crystalline phases is the easy alignment, which facilitates the processing of the organic semiconductor significantly [5]. Currently, the design of discotic mesogens providing good charge transport properties in a suitable temperature range is of major interest [8–10]. Among these materials liquid crystalline crown ethers are particularly promising providing a unique entry into supramolecular chemistry due to their known propensity for selective metal salt complexation [11–13]. Previously, we reported about bi-centred liquid crystalline crown ethers and the impact of the molecular flexibility and geometry on structure and electronic charge transport properties in their columnar phases [14]. The mesogens consisted of a central crown ether that connects two n-alkoxy substituted triphenylene cores [15,16]. We demonstrated that the structure and the electronic transport along the stacked triphenylenes are strongly affected by the above-mentioned molecular parameters. The mesogen carrying a small rigid 12-crown-4 connecting the two triphenylenes as symmetric as possible led to the formation of a broad mesophase with high intra-columnar order, which provided the highest hole mobility among the investigated compounds [14]. Furthermore, complexation of 12-crown-4 and 18-crown-6 with Li^+ and K^+ salts, respectively, carrying soft anions rather than hard halides resulted in significant broadening or even the induction of columnar mesophases [17]. Complementary ^{1}H and ^{13}C NMR experiments in solution [17,18] revealed that the use of soft polarizable anions, like iodide and thiocyanate, led to the formation of tight ion pairs in the case of the KI- and KSCN-complex of the 18-crown-6 derivative. These results suggested that Coulomb interactions between the mesogens should improve the columnar order.

Previously reported small angle X-ray diffraction data revealed that the broad columnar hexagonal phase (*p6mm*) of the neat 12-crown-4 derivatives became even wider after complexation with LiI and the lattice constant *a* increased [15]. On the other hand, the neat 18-crown-6 derivatives possessed a narrow columnar rectangular mesophase (*c2mm*) [18]. Due to the coordination of K^+ ions accompanied by soft anions the phase type changed. The KI and KBF_4 complexes displayed a highly ordered Col$_r$ phase with *p2mg* symmetry, while the KSCN complex showed a highly ordered Col$_r$ phase with *p2gg* symmetry.

Since the observed salt-induced changes in the phase behaviour must be related to the molecular arrangement of the columns, the stacking of the triphenylenes within the column should be affected as well. Thus, we surmised that ion uptake should not only change the structure, but also the electronic transport. This concept of combining aromatic groups for the electronic transport along the stacks with a crown ether that is able to coordinate specific types of cations leads to several possible features of the columnar phase. The coordination of ion pairs provides the unique opportunity to change the intra-columnar packing of the mesogens without varying any of the important remaining material parameters, like the aromatic cores or the lateral side chains (Figure 1).

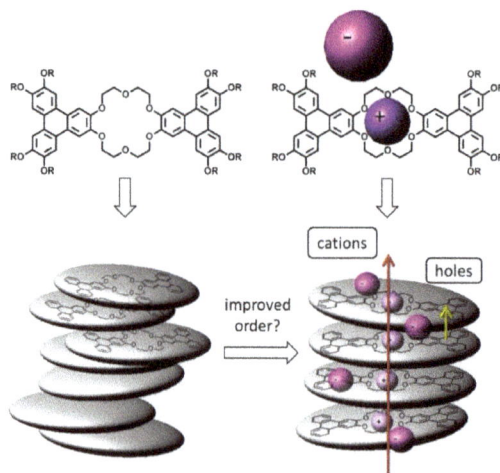

Figure 1. Structures of the bi-centred crown ether based discotic mesogens in this study and the expected impact of salt complexation. The coordinated ions are expected to improve the order and thus the hole transport via the stacked triphenylenes. Furthermore, the mesophase should provide channels for fast cation transport along the columns due to the stacked crown ether moieties.

Furthermore, we anticipated that, most likely, anisotropic ionic conductivity might be induced in the columnar phase. Since the crown ether moieties are stacked on top of each other in the centre of the columns channels for fast cation transport might be generated in the mesophase [19]. For smectic liquid crystal phases, strongly anisotropic two-dimensional (2D) ionic conductivity was demonstrated [20]. The transport of Li$^+$ in smectic phases of PEO-based (poly(ethylene)oxide) mesogens [21,22] is up to 100 times faster parallel to the smectic layers than perpendicular to them. Similar behavior was found for H$^+$ in smectic phases of mesogens with terminal diol-moieties [23]. Additionally, in columnar phases, strongly anisotropic 1D ion transport through the column centre was already observed by Yoshio et al.in columnar phases of an imidazolium-based ionic liquid crystal [24]. Beginn et al. could show that the membranes of polymerized columnar phases of liquid crystalline crown ethers where 4–5 molecules form supramolecular discs are at least permeable for several ions [25,26].

In the current study, the validity of the concept outlined in Figure 1 was probed by investigation of the electronic and ionic transport in the mesophase of the Li$^+$-complex of 12-crown-4 and the K$^+$-complex of 18-crown-6 based discotic mesogens. The transport properties were studied by photoconductivity (electronic transport) and impedance spectroscopy (ionic transport), respectively, and they were correlated to structural observations from X-ray experiments. As described below, our results reveal that the enhanced intra-columnar order in the KSCN and KI complexes of the 18-crown-6 derivative leads to improved photoconductivity, which is mainly due to anion mobility rather than cation transport.

2. Materials and Methods

2.1. Synthesis of the Crown Ether Salt Complexes

LiI complexes of 12-crown-4 **LiI-1a** (R = C$_9$H$_{19}$) and **LiI-1b** (R = C$_{12}$H$_{25}$) were prepared according to our previously published procedure [15]. KX complexes of 18-crown-6 **KI-2a** (R = C$_9$H$_{19}$) and **KI-2b**, **KSCN-2b** (R = C$_{11}$H$_{23}$), and **KI-2c**, **KSCN-2c**, and **KBF$_4$-2c** (R = C$_{12}$H$_{25}$) were prepared according to our previously reported method [18]. The complexation of the known crown ethers **1a,b** and **2a–c** was performed by adding a solution of the crown ether in CH$_2$Cl$_2$ to a solution of 1.5 equiv. of the respective LiI or KX salt in MeOH, stirring for 18 h at room temperature, followed by filtration, and then evaporation of the solvent. As described in refs. [16–18], the formation of (1:1) complexes was monitored by the characteristic ^{1}H and ^{13}C NMR chemical shifts and MALDI-TOF MS spectra.

2.2. X-Ray Diffraction

Small angle X-ray scattering experiments were performed on a SAXSess system (Anton Paar, Graz, Austria), equipped with an advanced collimation block providing a very narrow line shaped X-ray beam. The X-ray source is a Cu-K$_\alpha$ X-ray tube providing a monochromatic wavelength of 0.1542 nm. The detector is either a CCD camera with a pixel size of 24 × 24 µm^2 or a 5 × 20 cm imaging plate read out in a imaging plate reader (Perkin Elmer Cyclone plus, Waltham, MA, USA). The sample is contained in quartz capillaries, with a diameter of 0.7 mm being placed in a temperature controlled sample holder (25–300 °C). The accessible q-range is 0.04–27 nm^{-1}.

The wide angle X-ray scattering experiments have been performed on a home-made imaging plate camera with a sample to imaging plate distance of 10 cm. The source is a Cu-K$_\alpha$ X-ray tube (Siemens Kristalloflex X-ray generator, Erlangen, Germany), providing a monochromatic wavelength of 0.1542 nm. The sample is placed in a small hole (about 2 mm in diameter) in a brass block as thin film and was kept between two permanent magnets providing a magnetic field of about 2 Tesla. The imaging plate is developed in an imaging plate reader (Fujifilm BAS SR, Tokyo, Japan). A Lakeshore 331 controller varies temperature.

To avoid the precipitations of the salt the compounds were filled into the capillary or the hole as a powder. During the X-ray experiments, the samples were never heated above their clearing point.

2.3. Photoconductivity

The sample was contained in typical liquid crystal glass cells (0.8 and 1.3 μm gap) equipped with ITO (Indium Tin Oxide) electrodes and rubbed polyimide alignment layers on both sides. Cells were filled with the isotropic melt of the compounds by capillary action. The temperature of the sample was controlled by a hot stage (Mettler Toledo FP-5, Columbus, OH, USA). A dc electric field of 0.25–1 V μm^{-1} is applied by a power supply. The sample was illuminated by a Xe-lamp (Perkin Elmer 150 W, Waltham, MA, USA) at a certain wavelength that was adjusted by an optical filter at (366 ± 5) nm or a monochromator (Horiba Spex 1681B, Kyoto, Japan). The light was chopped at the frequency of 6 Hz by a mechanical chopper, while a lock-in amplifier detected the photocurrent (Stanford Research Systems SR830 DSP, Sunnyvale, CA, USA). Data acquisition was done by a computer equipped with *LabView*.

2.4. UV/Vis Spectroscopy

Spectra were taken using a UV-Vis Spectrometer (Perkin Elmer Lambda 2, Waltham, MA, USA) with a wavelength range of 190–1100 nm. The samples are contained in 0.8 μm polyimide coated liquid crystal cells that were placed in a homemade temperature controlled sample holder.

2.5. Alignment and Polarizing Microscopy

The alignment of the sample and the changes in texture at the phase transitions have been tracked by a polarizing microscope (Olympus BH-2, Tokyo, Japan) equipped with a hot stage (Instec Mk2, Boulder, CO, USA). Large homeotropic aligned domains could be grown in parallel rubbed polyimide coated cells after careful thermal cycling close to the clearing point. All of the compounds showed a certain degree of macroscopic segregation between complex and neat compound after heating to the isotropic melt.

2.6. Impedance Spectroscopy

An impedance analyser (Hewlett Packard 4192 A, 5 Hz–13 MHz, Palo Alto, CA, USA), equipped with a homemade temperature controller was used. The amplitude of the ac-voltage was set to 0.2 V. The sample was measured using interdigitating platinum electrodes (d = 225 nm) on glass substrates (Schott AF45, Jena, Germany). The electrode structure provided a channel length of 5 μm and a total width of 115 cm. The ac-response of a liquid crystalline sample on interdigitating electrodes can be modelled by a simple *RQ*-equivalent circuit (Figure 2). *Q* is a constant phase element accounting for the slightly depressed semicircles in the *Nyquist* diagram.

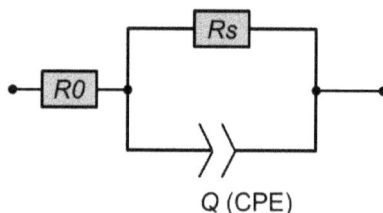

Figure 2. Equivalent circuit for the modeling of the impedance measurements. *R0* is the resistance of the wires and the electrodes themselves. *Rs* is the bulk dc-resistance of the sample and *Q* is the constant phase element (CPE).

The bulk resistance *Rs* was deduced from the diameter of the semicircle in the *Nyquist* diagram. The cell constant *A* in cm^{-1} was determined by calibrating the setup with 1 mM KCl-solution of known conductivity. The ionic dc-conductivity σ_{ion} was determined according to:

$$\sigma_{ion} = \frac{A}{R_s} \tag{1}$$

The liquid crystal was put on the surface covered by a thin glass slide and then heated to the columnar phase to equilibrate. The specific alignment of the columns was achieved by shearing the material in the liquid crystal phase either parallel or perpendicular to the electric field of the interdigitating electrodes. Data analysis was done using *ZView* (Scribner Associates Inc., Southern Pines, NC, USA).

3. Results and Discussion

3.1. Mesophase Structure of the Complexes

Figure 3 shows the investigated liquid crystalline crown ethers. The well-defined cavity size of the crown is responsible for the coordination of specific types of cations, which leads to the formation of (1:1) complexes with several ion pairs in solution.

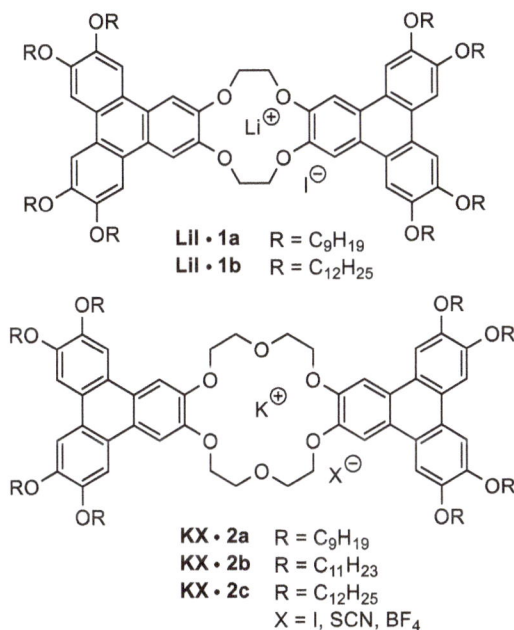

Figure 3. Investigated LiI-complexes of the 12-crown-4 and KX-complexes (X = I, SCN, BF_4) of the 18-crown-6 derivative bearing lateral side chains of OC_9H_{19}, $OC_{11}H_{23}$, and $OC_{12}H_{25}$.

For the detailed determination of the mesophase structure and the analysis of the electronic and ionic transport properties the alignment of the materials is of crucial importance. After filling the isotropic melt of the complexes into polyimide coated liquid crystal cells, it turned out that a significant part of the coordinated salt was macroscopically precipitated. The conglomerate was visible in the light microscope. This led to a phase separation of the free crown ether and the complex. Figure 4 shows the polarizing micrographs of the columnar phase of **KSCN-2b** in a liquid crystal cell after cooling down from the isotropic melt. Above 132 °C, dark parts are visible that are due to isotropic neat crown ether **2b** (Figure 4a). Upon further cooling (top to bottom) the isotropic-to-columnar phase transition of neat **2b** at 132 °C was clearly visible (Figure 4b).

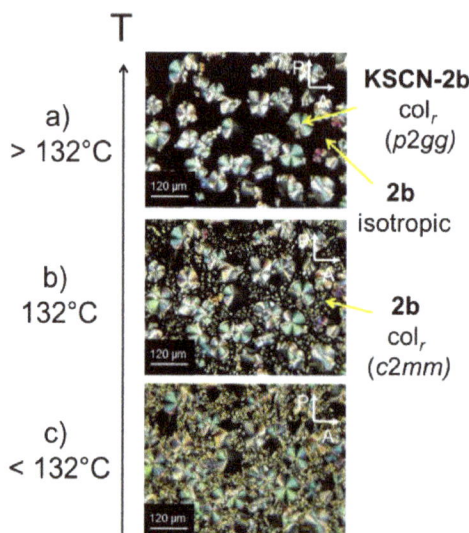

Figure 4. KSCN-2b sample in a polyimide coated cell after cooling down from the isotropic melt. Above 132 °C (**a**) the macroscopic segregation of the complex (texture) and isotropic free crown **2b** (dark parts) can be seen clearly. By cooling down (top to bottom) the isotropic-columnar phase transition (**b**) of **2b** at 132 °C can be observed. In (**c**), the columnar phase of **KSCN-2b** and neat **2b** coexist.

Approximately half of the crown ether complexes in the sample kept their ion pairs and they were able to self-assemble into the columnar mesophase even after repetitive heating/cooling cycles. For the 18-crown-6 derived K^+ complexes the tendency towards thermal decomplexation was dependent on the counterions. Complexes with SCN^- were more stable than those with I^-, BF_4^-, Br^-, and Cl^- counterions. In case of Cl^- or Br^-, the respective salt KCl or KBr precipitated already partially in the columnar phase and completely after heating to the isotropic phase. Due to the problems that are caused by the thermal decomplexation, we focused on the 18-crown-6 complexes **KI-2b** and **KSCN-2b** and the corresponding 12-crown-4 complex **LiI-1b** and re-examined them by SAXS (small-angle X-ray scattering).

The known diffraction patterns for **KI-2b** (Col$_r$, *p2mg*), **LiI-1b** (Col$_h$, *p6mm*), and **KSCN-2b** (Col$_r$, *p2gg*) [15–18] could be confirmed. However, for **KSCN-2b**, we discovered an additional 46 K wide high temperature phase with *p2mg* symmetry between 190 °C and the clearing point at 236 °C (Figure 5, Table S1). The lattice constants of the low temperature phase (*p2gg*) a = 53.1 Å and b = 47.4 Å (at 180 °C) are very similar to those found previously [15–18]. The lattice parameters of the new high temperature phase (*p2mg*) a = 64.1 Å and b = 38.2 Å (at 220 °C) are significantly different. Figure 5 shows the small angle diffraction patterns of both columnar phases and the solid state of a polydomain sample of **KSCN-2b**. The X-ray data with the respective Miller indices are summarized in Table S1. In both columnar phases, numerous sharp reflexes can be observed, indicating a high degree of inter-columnar order.

In the solid state of **KSCN-2b**, instead the diffraction pattern changed to diffuse scattering, where only one single broad peak plus a small shoulder could be clearly identified. At higher angles, several broad peaks were detected, indicating that the well-defined crystal structure of neat **2b** (see [14]) is turned into a glassy state *g* by the coordination of KSCN. The SAXS pattern in Figure 5 however indicates, that the inter-columnar order of the *p2gg* phase is not fully preserved in the glassy state, which might originate from conformational changes of the mesogens below the glass transition. The new phase sequence of **KSCN-2b** is shown in Figure 6. Since the Col$_r$-Col$_r$ phase transition could neither be seen via POM (polarizing optical microscopy) or DSC (differential scanning calorimetry) on cooling, the value is shown in parentheses.

Even though we do not have the full XRD data set for complex **KSCN-2c** due to lack of material, we assume that the phase behaviour is rather similar to **KSCN-2b**, since both of the complexes showed very similar behaviour in the DSC (Figure S2 in the Supplementary Materials) as well as in the POM.

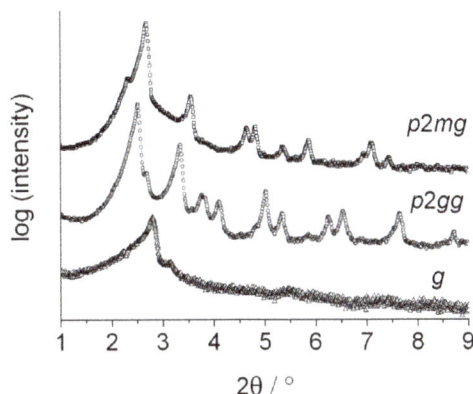

Figure 5. Small angle diffraction pattern of the new high temperature Col$_r$-phase (*p2mg*) (□) at 220 °C, the already known col$_r$-phase (*p2gg*) (○) at 180 °C and the solid state (*g*) at 20 °C (△) of **KSCN-2b**.

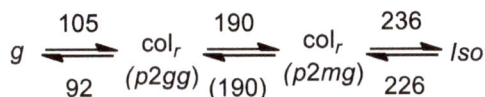

Figure 6. New phase sequence of **KSCN-2b**. The phase sequence of **KSCN-2c** is assumed to be very similar.

After having clarified the mesophase geometries, we wondered how the coordination of ions might affect the intra-columnar order and thus the electronic properties of the material. Therefore, we performed detailed wide angle X-ray diffraction experiments. Figure 7 exemplarily shows the comparison of the wide angle diffraction patterns of neat crown ethers **1b**, **2c**, and their respective LiI- and KSCN-complexes.

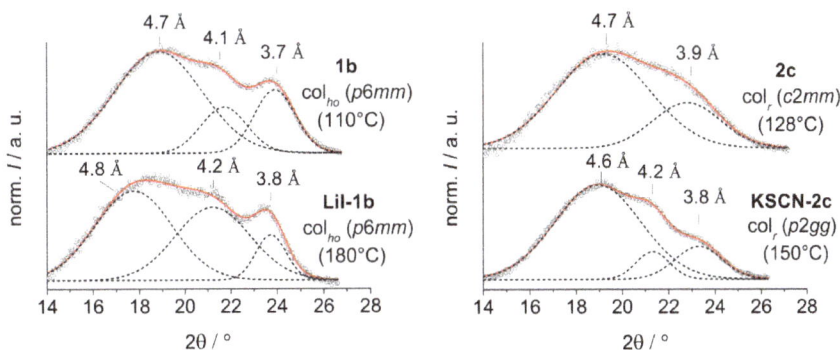

Figure 7. Wide angle diffraction patterns of **1b** vs. **LiI-1b** (**left**) and **2c** vs. **KSCN-2c** (**right**). An additional peak can be observed in the diffraction pattern of **KSCN-2c** due to the ion coordination. The dotted curves divide the scattering pattern in single Gaussian distribution functions.

The XRD patterns of the 12-crown-4 derivatives **1b** and **LiI-1b** were very similar. In each case, three peaks could be observed. The broad halo at lowest angle is attributed to the liquid, like alkyl chains with a mean distance of about 4.7 Å. The peak at highest angle belongs to the stacking of the triphenylene cores with a mean distance of 3.7 Å. The peak in the middle is probably due to the periodic arrangement of the small rigid crown ethers. The distance between the crown ether units is about 4.1–4.2 Å. The scattering of the triphenylenes indicated a slightly increased correlation length (lower peak width) in the case of **LiI-1b** as compared to **1b**.

The XRD patterns of the 18-crown-6 derivatives showed clear differences. While the scattering of the crown ether units cannot be observed in neat **2c** the peak was visible in the complex **KSCN-2c**. Furthermore, the stacking period of the triphenylenes decreased from 3.9 to 3.8 Å. Presumably, due to the coordination of K^+, the flexibility of the crown ether is restricted by the coordinative bonds of the oxygen atoms to the central cation.

This model is in good agreement with previous studies on crystal structures of salt complexes of aryl-substituted crown ethers, which revealed that oxygen atoms are arranged in a plane surrounding the cation and the aromatic residues are oriented out of this plane, leading to a bowl-shaped arrangement [27–30]. Such planar arrangement should result in decreased intracolumnar distances between neighboring mesogens. The increased order of the oxygen atoms leads to the additional diffraction peak at 4.2 Å, which was observed only for the small rigid 12-crown-4 derivatives **1**.

These results show that the coordination of ions promotes a higher intra-columnar order. The strong broadening of the columnar phases can be explained by these findings. The increased order and the decreased stacking distance of the triphenylenes should improve the electronic transport along the stacks.

In the solid state the complexes **LiI-1b** and **KSCN-2c** showed the expected broad diffraction patterns bearing only diffuse peaks (Figure S4), which are typical for a glassy state. Since the intra-columnar order is very low, the electronic charge transport should be disfavoured in the solid states of the complexes.

3.2. Electronic Transport

The combination of photoconductivity measurements and impedance spectroscopy allows for the complementary detection of either the electronic or the ionic transport in the columnar material. The use of these two different methods provides insight into the electronic properties of mixed conductors. We chose the lock-in technique to measure the photoconductivity, because it is a useful tool to separate the electronic from the ionic transport. For a good charge transport along the stacked triphenylene columns, homeotropic alignment in the liquid crystal cell is required. The alignment was achieved by surface interaction of the mesogens at the isotropic-to-columnar phase transition. For the subsequent experiments, **KSCN-2c** was used, because it showed sufficient stability during repetitive heating/cooling cycles and the growth of large homeotropically aligned domains of the complex on the electrode area. According to our previous results, neither the crown ether size nor the length of the side chains influences the absorbance spectra of the compounds [14]. In current experiments, we observed that the presence of coordinated salts in the crown does not change the shape and position of the absorption bands as exemplified for **2b** and **KSCN-2b** (Figure S2). Since the liquid crystal cells are fabricated from glass blocking wavelengths shorter than 310 nm, the only suitable optical transitions are the absorption bands at 346 and 363 nm belonging to the vibration fine structure of the $S_0 \rightarrow S_1$ transition of the triphenylene unit [31].

Figure 8 shows the typical temperature dependent photoconductivity profile of **KSCN-2c** (red curve) without any contributions from the segregated **2c** due to its edge-on (planar) alignment. The measurement was obtained using a 3.5 µm polyimide coated cell that was illuminated at 370 nm via monochromator. The applied electric field was held at 1 V µm^{-1}. The second graph shows the photoconductivity profile of neat **2c** (black curve). The photoconductivity of **KSCN-2c** displayed a strong dependence on the phase type and temperature. As expected, in the isotropic state, the photocurrent is very close to zero. Both Col$_r$ mesophases (*p2mg* and *p2gg*) instead provided significant photoconductivity. The maximum response was detected at about 160 °C in the low temperature mesophase with *p2gg* symmetry. The glassy solid state *g* appears to be completely insulating regarding the electronic transport. All of the phase transitions were visible in the profile, but they are not as distinct as in the corresponding photoconductivity profile of the neat crown ether **2c**, making the assignment of a sharp transition temperature difficult. The highest signal for the neat crown ether **2c** was measured in the crystalline states Cr$_1$, Cr$_2$ at about 65 °C, while the rectangular columnar phase (*c2mm*) between 123 and 133 °C showed considerably lower photoconductivity. All of the phase transitions were very distinct in the profile of neat **2c**. It is evident from Figure 9 that neat 18-crown-6 **2c** and the complex **KSCN-2c** display complementary behaviour. The electronic transport in the crystalline solid state of **2c** was totally suppressed by ion uptake. In contrast, in the columnar phase of neat 18-crown-6 **2c**, no electronic transport was detected, whereas ion uptake, i.e., the formation of the complex **KSCN-2c** resulted in a significant increase of the electronic transport as compared to the glassy solid state.

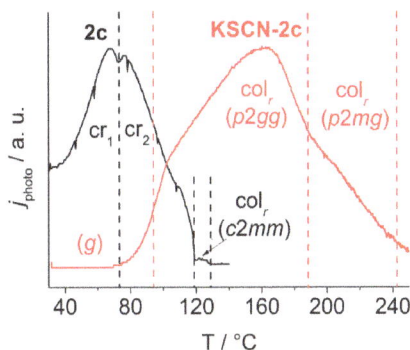

Figure 8. Typical normalized temperature dependent photoconductivity profile of **KSCN-2c** (red) and neat **2c** (black). The phase transitions are depicted by the color coded vertical dashed lines for **2c** and **KSCN-2c**. Neat **2c** shows low photocurrent in the columnar phase (*c2mm*) and a significantly raised signal in both crystalline solid states Cr$_1$ and Cr$_2$. The complex instead shows an "inverted" profile. The two columnar phases (*p2gg* and *p2mg*) provide significant photoconductivity while the signal vanishes completely in the glassy solid *g*. In both cases, no current could be detected in the isotropic melt.

To achieve optimum conditions for quantitative comparison of the photoconductivity in the columnar phases of both compounds, neat 18-crown-6 **2c** and **KSCN-2c** were aligned in a 0.8 μm polyimide-coated cell. The applied electric field was held at 0.25 V μm^{-1} and both light intensity and wavelength were controlled by an optical filter at (366 ± 5) nm. The measurement was performed between the clearing point of the respective compound and room temperature at a cooling rate of 2 K min^{-1}. Figure 9a) shows the comparison of the measured photocurrent j_{photo} in the Col$_r$ phases of neat **2c** and **KSCN-2c**. The homeotropic alignment of neat **2c** and **KSCN-2c**, respectively, on the electrode area in the cell are shown in Figure 9b,c.

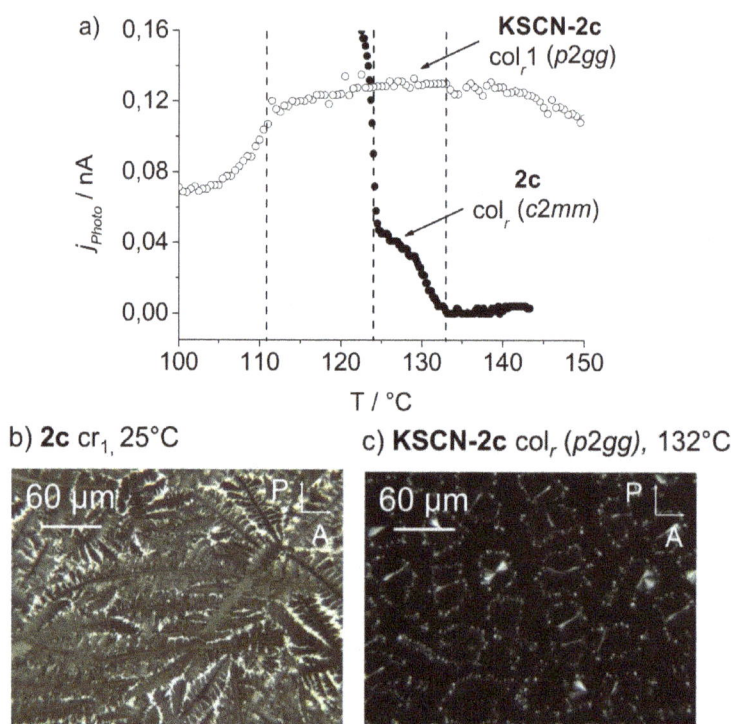

Figure 9. Temperature dependent photoconductivity (**a**) of the columnar phase of neat **2c** (●) and the low temperature columnar phase of **KSCN-2c** (○) both contained in a 0.8 μm LC (liquid crystal) cell. The polarizing micrographs show the nearly homeotropically aligned neat **2c** (**b**) all over the electrode area, in contrast to the phase separated **KSCN-2c** in the liquid crystal cell (**c**).

It should be emphasized that a quantification of the experimentally determined photocurrents has to be considered with great care for several reasons. Only about 50% of the electrode area in the **KSCN-2c** sample was covered with the homeotropically aligned mesophase of the complex, while the remaining area was partly covered with free 18-crown-6 derivative **2c**, which is isotropic at temperatures exceeding 120 °C and thus does not contribute to the total photoconductivity. In the sample of neat **2c**, about 90% of the electrode area was covered (Figure 9). This means that the effective intersection of the light beam with the photoactive parts of **KSCN-2c** is only about half as large as it is in the sample of neat **2c**. Additionally, a weakening of the internal dc-field in **KSCN-2c** due to the formation of electrolytic double-layers at the electrode surface in the presence of ions should be expected. This is not the case for neat **2c**. Despite these limitations, the results in Figure 10 suggest that the photoconductivity in the columnar phase of **KSCN-2c** is about three times higher as compared to the neat 18-crown-6 **2c**, which is in agreement with the changes of the inter- and intracolumnar structure, as determined by SAXS and WAXS (wide-angle X-ray scattering) experiments. Unfortunately, the exact determination of the charge carrier mobility with the organic field effect transistor (OFET) was not possible for the complexes. The desired field effect current was strongly superimposed by the ionic current flow.

a) **KI-2c** columns ∥ $\vec{E}$

b) **KI-2c** columns ⊥ $\vec{E}$

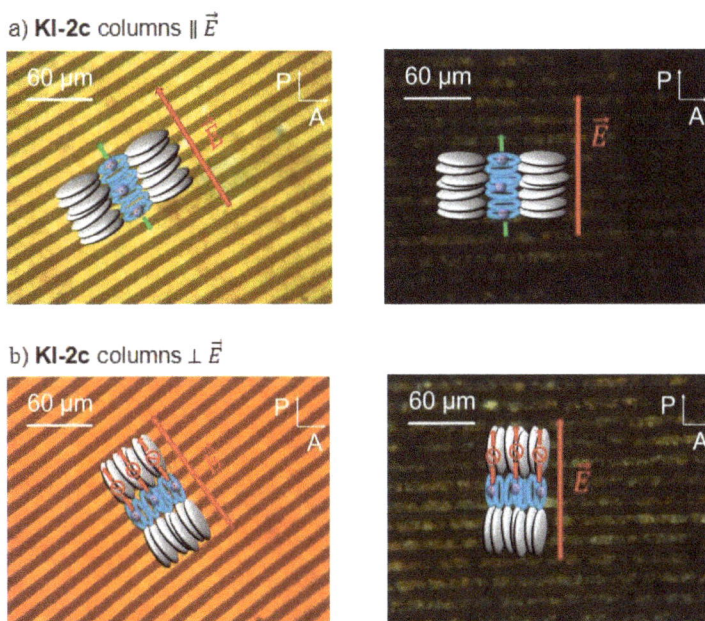

Figure 10. Bright and dark state polarizing micrographs of KI-2c films on interdigitating electrodes with columns aligned parallel to the electric field direction (**a**) and columns aligned perpendicular to the electric field direction (**b**).

3.3. Ionic Transport

To investigate the ionic conductivity, we performed impedance spectroscopy, which is a powerful tool to measure the bulk ionic dc-conductivity. Since the use of liquid crystal cells for these measurements was impossible due to the precipitation of the coordinated salt we used interdigitating gold electrodes. The liquid crystalline material was held at temperatures that were only slightly above the melting point during sample preparation.

In order to clarify, whether coordinated cations in the centre of each mesogen could migrate through the stacked crown ether moieties as 1D transport channels, experiments were carried out on planar aligned films with columns either parallel or perpendicular to the electric field direction. Figure 10 shows two aligned films of **KI-2c** in the liquid crystal phase at 140 °C in the bright and dark state between crossed polarizers. Since the photographs have been taken in reflexion, the bright stripes in the bright states are the platinum electrodes. In Figure 10a, the columns of the LC phase are aligned parallel to the electric field, while in Figure 10b the columns are aligned perpendicular.

By comparing the interference colour of the films in linear polarized light with samples of known thickness in liquid crystal cells, the thickness of the films could be estimated to be about 5 μm. Van Gerwen et al. calculated the electric field distribution in the case of interdigitating electrodes [32]. Applying this method to our geometry, where the 230 nm high and 5 μm wide electrode digits are separated by 5 μm, a minimum film thickness of 10 μm is required to cover 100% of the electric stray field. On the other hand, a 5 μm thick film still covers about 80% and the thickness of both films is quite similar (yellow/orange main colour). This means that the comparison of the measured dc-conductivity values for the films should be reliable.

Figure 11 shows the results for **KI-2c** measured parallel to the columns at 150 °C in the LC-phase. The frequency dependence of the total impedance $|Z|(\omega)$ and the phase angle $\phi(\omega)$ are displayed in the *Bode* diagram. Furthermore, the *Nyquist* diagram $-Z''(Z')$ is shown where the imaginary part of the impedance is plotted versus the real part, leading to a semicircle.

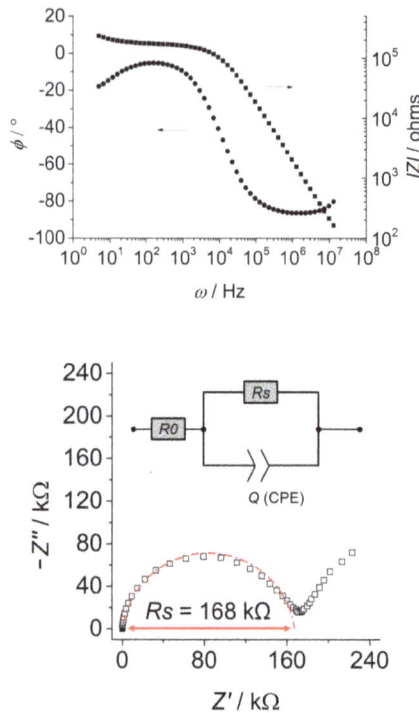

Figure 11. *Bode* plot showing the frequency dependence of the total impedance $|Z|(\omega)$ and the phase angle $\phi(\omega)$ (top) and the *Nyquist* plot $-Z''(Z')$ (imaginary part vs. real part of the impedance) showing a slightly depressed semicircle that was fitted by the equivalent circuit shown in the inlet (bottom).

The equivalent circuit we used for fitting the data is shown in the inlet. The pre-resistance of the wires and the platinum electrodes themselves is represented by R0. It is in series with the standard RQ-circuit that was used for the modelling of electrolytes on interdigitating electrode structures. The constant phase element Q accounts for the slightly depressed shape of the semicircle. The diameter of the semicircle is given by the bulk dc-resistance Rs of the film. With the specific cell constant A of $0.0076\ cm^{-1}$, the ionic dc-conductivity is determined from Equation (2) (see below). At low frequencies, the onset of a second semicircle could be observed in the *Nyquist* diagram. This phenomenon could be due to the formation of an ionic double layer at the electrode surface. Another possibility would be the contribution of a second transport process, for example, the migration through domain boundaries. In crystalline solids, the strong effect of grain boundaries on the charge transport is well known [33]. Since the low frequency arc did not affect the measurement results, its origin was not investigated further. Figure 12 shows the ionic conductivity parallel $\sigma_{\parallel}$ and perpendicular $\sigma_{\perp}$ to the columns determined from the measurements of the aligned films of **KI-2c** shown in Figure 10. Both of the films were measured between 120 and 155 °C in the Col$_r$ phase (*p2gg*).

It is clearly seen that $\sigma_{\parallel}$ is slightly higher than $\sigma_{\perp}$, but the maximum anisotropy at 155 °C (= 2.33 × 10^{-3}/K in Figure 12) is only $\sigma_{\parallel}/\sigma_{\perp} = 1.7$. If channels for fast K$^+$ transport exist in the columnar phase, the value of $\sigma_{\parallel}$ should be much higher than $\sigma_{\perp}$. Parallel to the columns, the cations should migrate through the stacked crown ethers but perpendicular to the columns the cations cannot contribute to the conductivity, since they would have to leave their coordination site (compare the inset of Figure 10).

In comparison to liquid crystalline materials providing channels for fast ion transport with anisotropy values between 10 and 100 [21,23,24], the observed difference in conductivity is most probable not due to the 1D transport of the coordinated K^+ cations. Furthermore, the measured conductivity values turned out to be rather low never exceeding 10^{-7} S cm^{-1}, which is not typical for materials featuring ion channels. The measured anisotropy was in the order of magnitude of a conventional nematic liquid crystal where the anisotropy in ionic conductivity is just caused by the anisotropic structure of the liquid crystal, as shown by Stegemeyer et al. [34]. Presumably, the K^+ ions are strongly coordinated and the ion migration should be dominated by the non-coordinated anions that migrate through the side chains of the columnar system. Since the transport seems not to be dependent on the orientation of the columns in the liquid crystal phase and other investigated compounds showed anisotropy values ≤1.7, the following measurements were carried out on randomly aligned thicker films (>10 μm).

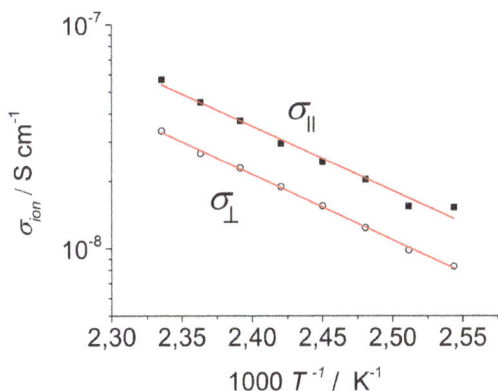

Figure 12. Arrhenius diagram of the ionic conductivity parallel $\sigma_{\parallel}$ (■) and perpendicular $\sigma_{\perp}$ (○) to the columns in the columnar phase of **KI 2c**.

Since the anion seems to play the decisive role, we investigated the K^+ complexes **KX-2c** with different anions (X = I, SCN and BF$_4$) (Figure 13). Upon using smaller counter ions, like Br$^-$ or Cl$^-$, the precipitation of the respective salt already took place at the solid-to-columnar phase transition, making it impossible to get reproducible conductivity values.

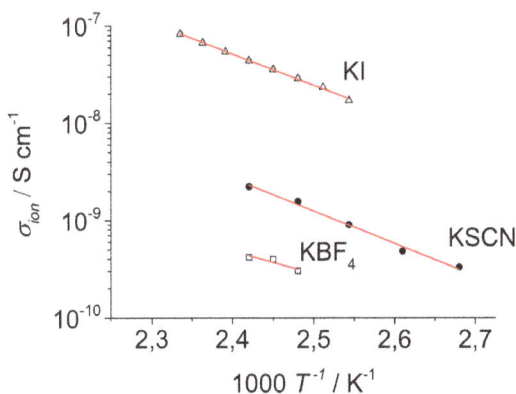

Figure 13. Arrhenius diagram of the ionic conductivity of **KI-2c** (Δ), **KSCN-2c** (●), and **KBF₄-2c** (□).

The conductivity for **KI-2c** (4×10^{-8} S cm^{-1} at 140 °C = 2.42×10^{-3}/K in Figure 13) was 20 times higher than for **KSCN-2c** and 100 times higher than for **KBF$_4$-2c**. This result suggests that the anions dominate the contribution to the ionic conductivity. This result might be rationalized by SCN$^-$ acting as a bridging ligand. If the SCN$^-$ is coordinating two K$^+$ complexes, then it would be firmly incorporated in the columnar structure, which would be detrimental to its mobility in the electric field. On the contrary, the less polarizable BF$_4$$^-$ might be less soluble in the liquid crystal and precipitation of KBF$_4$ might cause a lower charge carrier concentration and thus lower conductivity.

Thus, different kinds of counter ions have a strong impact on the ionic conductivity of the columnar phase. Since the melting points were in the same temperature range (125 °C for **KI-2c**, **KBF$_4$-2c** and 110 °C for **KSCN-2c**) and all of the compounds form columnar rectangular mesophases, a similar viscosity was proposed for these derivatives. Among the tested anions, I$^-$ is traveling much faster than SCN$^-$ or BF$_4$$^-$, resulting in more than one order of magnitude higher conductivity. These results indicate comparatively strong bound cations in the centre of the crown ether moiety of each mesogen plus mobile anions that are located mainly in the peripheral liquid, like side chains. The observed anisotropy of $\sigma_\parallel/\sigma_\perp = 1.7$ (at 155 °C) could be explained by considering the local viscosity anisotropy of the side chains, since the viscosity scales with the order parameter of the alkyl chains. This might indicate enhanced conductivity perpendicular to the columns, which was not observed. On the other hand, ions migrating along the columns do not have to overcome the barriers of stacked crown ether moieties. Thus, ion migration along the columns might be favorable for the anions.

Since the anions seem to be mobile parallel and perpendicular to the columns (Figure 12), we assume that they are located mainly in the amorphous side chains of the columnar system. The viscosity η of the solvent influences the ion mobility μ_{ion}, according to the Stokes–Einstein equation:

$$\mu_{ion} = \frac{ze}{6\pi\eta r_{ion}} \tag{2}$$

where z describes the charge number, e the elementary charge, and r_{ion} the radius of the solvated ion. By the variation of the side chains it should be possible to change the environment of the anions specifically, since the cations are located in the centre of the columns. By elongation of the side chains, the local viscosity should be lowered, which would result in higher ion mobility according to Equation (2). A similar effect is expected by enlarging the central crown ether moiety. Therefore, we compared thick films (>10 μm) of the iodide-complexes with different chain lengths (-OC$_9$H$_{19}$ or -OC$_{12}$H$_{25}$), namely **LiI-1a,b** and **KI-2a,c** (Figure 14).

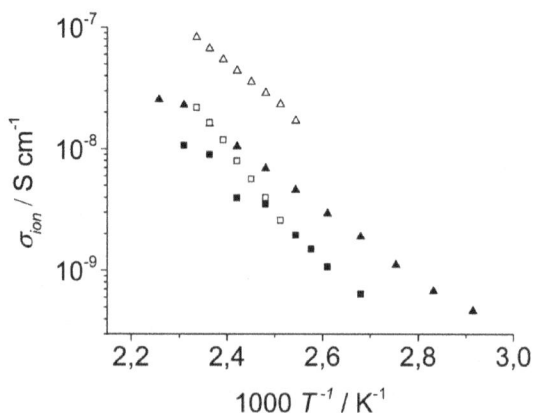

Figure 14. Arrhenius diagram of the ionic conductivity of **LiI-1a** (■), **LiI-1b** (▲) **KI-2a** (□), and **KI-2c** (△). Both pairs of compounds with neat and filled symbols, respectively, bear the same ion pairs but they differ in side chain length.

Upon comparison of the homologues with different side chain length, the derivatives with C12 chains (▲ and △) showed higher conductivity than those with C9 chains (■ and □). When comparing the K$^+$ complexes of the 18-crown-6 derivatives with the Li$^+$ complexes of the 12-crown-4 derivatives with identical side chains (△ vs. ▲ and □ vs. ■) the higher conductivity was observed for the complexes **KI-2a,b**. The ion mobility can be increased by elongated side chains. Since the side chains should not affect the transport of the coordinated cations in the column centres, the results in Figure 14 suggest that the anions dominate the conductivity of the columnar phase and that they are located mainly in the amorphous side chains.

The observed anisotropy of $\sigma_{\parallel}/\sigma_{\perp} = 1.7$ favors the migration of the anions parallel to the columns. This might be rationalized by comparing these columnar systems with a nematic liquid crystal. In a conventional nematic liquid crystal, the transport of ions is favored parallel to the director, i.e., parallel to the long axis of the rod-like molecules. A value of the anisotropy $\sigma_{\parallel}/\sigma_{\perp} = 1.3$ was published by Stegemeyer et al. [34], which is somehow comparable to our case, if the columns of the columnar phases of the crown ether complexes are considered as rods of an ordered nematic phase.

When considering that only the anion contributes to the charge transport and that every mesogen incorporates one ion pair, it is also possible to derive mobility values for the different anions. The specific conductivity of electrolytes σ is given by:

$$\sigma = \Lambda_m c, \tag{3}$$

with Λ_m being the molar conductivity and c the concentration of charge carriers. Excluding the migration of cations, the molar conductivity can be calculated as:

$$\Lambda_m = |z_-|v_- F \mu_-, \tag{4}$$

where $|z_-|v_-$ is the electrovalence of the anions being 1 in our case, F is the Faraday-constant, and μ_- is the mobility of the anions. Since there is one anion per mesogen, the concentration can be estimated by the density of mesogens per unit volume in the columnar phase. Finally, the mobility of the anions can be derived from:

$$\mu_- = \frac{\sigma}{F} \frac{M}{\varrho}, \tag{5}$$

with M as molar mass of the mesogen and ϱ the density of the phase.

While the density could not be experimentally measured, the crystallographic density of the crown ethers substituted with aromatic units was taken from the literature 1.3 g cm^{-3} [27–30]. The coordination of ions increases the density, while the substitution with long alkyl tails in turn decreases it. Therefore, the density of 1.3 g cm^{-3} should be appropriate. According to these estimations, the mobility of the anions was determined at 140 °C for the different compounds and they are summarized in Table 1. When comparing these values with the recently determined hole mobility of the Col$_r$ phase of the neat **2c** [14], it can be seen that the hole transport in our materials is about four orders of magnitude faster than the anion transport.

Table 1. Calculated anion mobility in the columnar rectangular mesophases of **KI-2c**, **KSCN-2c**, and **KBF$_4$-2c** at 140 °C for the estimated density of 1.3 g cm^{-3}.

	Structure of the Phase	σ [S cm^{-1}]	M [g mol^{-1}]	μ_{anion} [cm^2 V^{-1} s^{-1}]
KI-2c	col$_r$ (*p2mg*)	2.0×10^{-8}	2301.3	3.6×10^{-10}
KSCN-2c	col$_r$ (*p2gg*)	2.2×10^{-9}	2232.5	4.0×10^{-11}
KBF$_4$-2c	col$_r$ (*p2mg*)	4.2×10^{-10}	2261.2	7.5×10^{-12}

4. Conclusions

The ion complexes of different bi-centred liquid crystalline crown ethers that are based on 18-crown-6 and 12-crown-4 moieties were studied regarding a correlation between the structure of the mesophases and their electronic and ionic transport properties. XRD experiments revealed that the free 12-crown-4 **1b** and its complex **LiI-1b** both exhibit a broad columnar hexagonal mesophase. No significant differences were found regarding the intra-columnar order. In contrast, the free 18-crown-6 **2c** exhibited a columnar rectangular mesophase ($\Delta T = 13$ K) with *c2mm* symmetry. Upon coordination of K^+ salts with soft anions, like I^- or SCN^-, the mesophase range was highly broadened and the symmetry of the phases changed. **KI-2c** possessed a Col_r phase ($\Delta T = 80$ K) with *p2mg* symmetry. **KSCN-2c** displayed a high temperature Col_r phase ($\Delta T = 46$ K) with *p2mg* symmetry and a low temperature Col_r phase ($\Delta T = 85$ K) with *p2gg* symmetry.

Furthermore, **KSCN-2c** showed an improved intra-columnar order, leading to a better packing of the crown ether moieties plus a shortened stacking distance of the triphenylene cores of 3.8 Å in contrast to 3.9 Å for the neat compound **2c**.

The increased intra-columnar order had a pronounced impact on the photoconductivity where the detected signal was about three times higher in the columnar phase of **KSCN-2c** than in neat **2c**.

Investigation of the ion migration on macroscopically aligned thin films of **KI-2c** revealed that the dc-conductivity parallel to the columns in the liquid crystal phase was only slightly higher than perpendicular to them. The anisotropy was determined to be $\sigma_{\parallel}/\sigma_{\perp} = 1.7$ for **KI-2c** at 155 °C. The type of the anion (I^-, SCN^-, BF_4^-) had a strong impact on the conductivity of the K^+-complex of **2c**. **KI-2c** showed 20 times higher conductivity than **KSCN-2b** and conductivity that was 100 times higher than **KBF$_4$-2c**. From the dependence of the ionic conductivity on the size of the central crown and the length of the lateral side chains in **LiI-1a,b** and **KI-2a,b**, we concluded that the ion migration is dominated by the non-coordinated anions propagating through the anisotropic liquid, like side chains, while the cations are strongly bound in the centre of the columns. Thus, the existence of channels for fast cation transport could be excluded. It should be noted that Bardaj, Espinet, and coworkers recently reported similar ionic conductivities for K^+ complexes of diaza-18-crown-6 ethers carrying six decyloxy-p-cyanobiphenyl chains and showing nematic mesophases [35]. The corresponding Li^+ complexes showed 10 times higher conductivity, which was rationalized by cations that were jumping from one crown to another close to it [35].

As showcased for 18-crown-6 **2c** and the corresponding complexes **KX-2c** with different anions, the ion complexes form highly ordered columnar mesophases with improved stacking of the triphenylene cores, leading to increased 1D photoconductivity (Figure 15). The coordinated cations are strongly bound to the centre of the columns while the non-coordinated anions are mobile in the anisotropic liquid like side chains. The calculated mobility for **KX-2c** with different anions was in the order of 10^{-12}–10^{-10} cm^2 V^{-1} s^{-1}, which is four orders of magnitude lower than the hole mobility of the neat 18-crown-6 **2c**. This again correlates well with the results by Bardaj, Espinet, and coworkers, which observed 20 times higher conductivity, when the crown ether metal complex was doped with the neat "empty" crown [35].

In conclusion, the complexation of columnar liquid-crystalline crown ethers did not lead to fast ion migration, as might be expected by the presence of crown ether channels. On the other hand, however, the coordination of ions in the channels improved the intra-columnar packing of the crown ether rings and thereby enhanced the electronic charge transport, namely the hole mobility. Even without fast ion channels, the complexation of liquid-crystalline crown ethers turned out to be a promising tool for tailoring the charge transport properties in this class of materials.

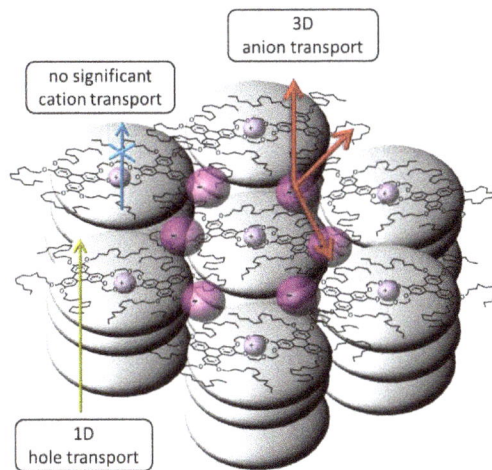

Figure 15. Our new mesogenic bi-centred ion complexes form highly ordered columnar mesophases providing improved quasi one-dimensional (1D) hole transport. The cations are coordinated strongly by the central crown ether moieties showing no significant contribution to the ionic conductivity. The anions are mobile in the anisotropic liquid like side chains.

Supplementary Materials: The following are available online at http://www.mdpi.com/2073-4352/9/2/74/s1. Figure S1: **KI-2b** in the columnar state (*p2mg*) (left) and in the solid state *g* (right). The texture shows no significant differences. Figure S2: DSC curves of the complexes **KSCN-2b** (a) and **KSCN-2c** (b). Heating/cooling rate 10 K/min. Figure S3: WAXS pattern of **KSCN-2b** taken from ref. [1]. Figure S4: Wide angle diffraction pattern of **LiI-1b** (○) and **KSCN-2c** (□) in the solid state. Only diffuse smeared peaks can be observed indicating a low intra-columnar order. Figure S5: Absorbance spectrum of neat **2b** (■) and its KSCN-complex **KSCN-2b** (●) in the solid state. Figure S6: Photoconductivity profiles of a segregated sample of **KSCN-2b** (△) where both components contribute to the photocurrent in the respective temperature ranges. Furthermore, the profile of neat **2b** (●) and a **KSCN-2b** sample (○) where only the complex contributes are shown. Figure S7: Large scale pictures of the aligned thin films of **KI-2b** between crossed polarizers in the bright state (left) and the dark state (right). The columns have been aligned parallel (A, B) and perpendicular (C, D) to the electric field. Figure S8: Temperature dependent ionic conductivity in thick films of **KI-2b** with columns aligned randomly (□), parallel (●) and perpendicular (▲) to the electric field. Table S1: Small angle X-ray data of **KSCN-2b**. θ is the scattering angle, *hk* the Miller indices, d_{ob} and d_{cal} are the observed and calculated distances while *a* and *b* are the lattice constants for the rectangular unit cell. Table S2: XRD data of **KSCN-2b** taken from ref. [1].

Author Contributions: Conceptualization, F.G. and S.L.; methodology, P.S. and M.K.; validation, P.S., M.K., F.G., S.L.; formal analysis, P.S., F.G., P.E., M.E.; investigation and synthesis, P.S., M.K.; data curation, P.S., F.G., S.L., P.E., M.E.; writing—original draft preparation, P.S., F.G., S.L.; writing—review and editing, S.L., P.E., M.E., F.G.; supervision, F.G., S.L.; project administration, F.G., S.L.; funding acquisition, F.G., S.L.

Funding: Generous financial support by the International Max Planck Research School for Advanced Materials (fellowships for P.S. and M.K.), the Deutsche Forschungsgemeinschaft (LA 907/17-1 SNAPSTER), the Bundesministerium für Bildung und Forschung (joint instrumentation grant) and the Ministerium für Wissenschaft, Forschung und Kunst des Landes Baden-Württemberg is gratefully acknowledged.

Conflicts of Interest: The authors declare no conflict of interest. The funders had no role in the design of the study; in the collection, analyses, or interpretation of data; in the writing of the manuscript, or in the decision to publish the results.

References

1. Adam, D.; Closs, F.; Frey, T.; Funhoff, D.; Haarer, D.; Schuhmacher, P.; Siemensmeyer, K. Transient photoconductivity in a discotic liquid crystal. *Phys. Rev. Lett.* **1993**, *70*, 457–460. [CrossRef] [PubMed]
2. Pisula, W.; Müllen, K. Discotic Liquid Crystals as Organic Semiconductors. In *Handbook of Liquid Crystals*, 2nd ed.; Goodby, J.W.G., Collings, P.J., Kato, T., Tschierske, C., Gleeson, H.F., Raynes, E.P., Eds.; Wiley-VCH: Weinheim, Germany, 2014; Volume 8, pp. 617–674.

3. Sergeyev, S.; Pisula, W.; Geerts, Y.H. Discotic liquid crystals: A new generation of organic semiconductors. *Chem. Soc. Rev.* **2007**, *36*, 1902–1929. [CrossRef] [PubMed]
4. Pisula, W.; Zorn, M.; Chang, J.Y.; Müllen, K.; Zentel, R. Liquid crystalline ordering and charge transport in semiconducting materials. *Macromol. Rapid Commun.* **2009**, *30*, 1179–1202. [CrossRef] [PubMed]
5. Pisula, W.; Menon, A.; Stepputat, M.; Lieberwirth, I.; Kolb, U.; Tracz, A.; Sirringhaus, H.; Pakula, T.; Müllen, K. A Zone-Casting Technique for Device Fabrication of Field-Effect Transistors Based on Discotic Hexa-*peri*-hexabenzocoronene. *Adv. Mater.* **2005**, *17*, 684–689. [CrossRef]
6. Seguy, I.; Destruel, P.; Bock, H. An all-columnar bilayer light-emitting diode. *Synth. Met.* **2000**, *111–112*, 15–18. [CrossRef]
7. Tang, C.W. Two-layer organic photovoltaic cell. *Appl. Phys. Lett.* **1986**, *48*, 183–185. [CrossRef]
8. Wöhrle, T.; Wurzbach, I.; Kirres, J.; Kostidou, A.; Kapernaum, N.; Litterscheidt, J.; Haenle, J.C.; Staffeld, P.; Baro, A.; Giesselmann, F.; et al. Discotic Liquid Crystals. *Chem. Rev.* **2016**, *116*, 1139–1241. [CrossRef]
9. Laschat, S.; Baro, A.; Steinke, N.; Giesselmann, F.; Hägele, C.; Scalia, G.; Judele, R.; Kapatsina, E.; Sauer, S.; Schreivogel, A.; et al. Discotic liquid crystals: From tailor-made synthesis to plastic electronics. *Angew. Chem. Int. Ed.* **2007**, *46*, 4832–4887. [CrossRef]
10. Eichhorn, H. Mesomorphic phthalocyanines, tetraazaporphyrins, porphyrins and triphenylenes as charge-transporting materials. *J. Porphyrins Phthalocyanines* **2000**, *4*, 88–102. [CrossRef]
11. Laschat, S.; Baro, A.; Wöhrle, T.; Kirres, J. Playing with nanosegregation in discotic crown ethers: From molecular design to OFETs, nanofibers and luminescent materials. *Liq. Cryst. Today* **2016**, *25*, 48–60. [CrossRef]
12. Kaller, M.; Baro, A.; Laschat, S. Liquid Crystalline Crown Ethers and Related Compounds. In *Handbook of Liquid Crystals*, 2nd ed.; Goodby, J.W.G., Collings, P.J., Kato, T., Tschierske, C., Gleeson, H.F., Raynes, E.P., Eds.; Wiley-VCH: Weinheim, Germany, 2014; Volume 6, pp. 335–376.
13. Kaller, M.; Laschat, S. Liquid crystalline crown ethers. *Top. Curr. Chem.* **2012**, *318*, 109–192. [PubMed]
14. Staffeld, P.; Kaller, M.; Beardsworth, S.J.; Tremel, K.; Ludwigs, S.; Laschat, S.; Giesselmann, F. Design of conductive crown ether based columnar liquid crystals: Impact of molecular flexibility and geometry. *J. Mater. Chem. C* **2013**, *1*, 892–901. [CrossRef]
15. Kaller, M.; Staffeld, P.; Haug, R.; Frey, W.; Giesselmann, F.; Laschat, S. Substituted crown ethers as central units in discotic liquid crystals: Effects of crown size and cation uptake. *Liq. Cryst.* **2011**, *38*, 531–553. [CrossRef]
16. Kaller, M.; Beardsworth, S.J.; Staffeld, P.; Tussetschläger, S.; Gießelmann, F.; Laschat, S. Increased mesophase range in liquid crystalline crown ethers via lower molecular symmetry. *Liq. Cryst.* **2012**, *39*, 607–618. [CrossRef]
17. Kaller, M.; Tussetschläger, S.; Fischer, P.; Deck, C.; Baro, A.; Giesselmann, F.; Laschat, S. Columnar mesophases controlled by counterions in potassium complexes of dibenzo18crown-6 derivatives. *Chem. Eur. J.* **2009**, *15*, 9530–9542. [CrossRef] [PubMed]
18. Kaller, M.; Deck, C.; Meister, A.; Hause, G.; Baro, A.; Laschat, S. Counterion effects on the columnar mesophases of triphenylene-substituted 18crown-6 ethers: Is flatter better? *Chem. Eur. J.* **2010**, *16*, 6326–6337. [CrossRef]
19. Yoshio, M.; Kato, T. Liquid Crystals as Ion Conductors. In *Handbook of Liquid Crystals, Second Completely Revised and Greatly Enlarged Edition*; Goodby, J.W.G., Collings, P.J., Kato, T., Tschierske, C., Gleeson, H.F., Raynes, E.P., Eds.; Wiley-VCH: Weinheim, Germany, 2014; Volume 8, pp. 727–750.
20. Funahashi, M.; Yasuda, T.; Kato, T. Liquid Crystalline Semiconductors: Oligothiophene and Related Materials. In *Handbook of Liquid Crystals*, 2nd ed.; Goodby, J.W.G., Collings, P.J., Kato, T., Tschierske, C., Gleeson, H.F., Raynes, E.P., Eds.; Wiley-VCH: Weinheim, Germany, 2014; Volume 8, pp. 675–708.
21. Hoshino, K.; Kanie, K.; Ohtake, T.; Mukai, T.; Yoshizawa, M.; Ujiie, S.; Ohno, H.; Kato, T. Ion-conductive liquid crystals: Formation of stable smectic semi-bilayers by the introduction of perfluoroalkyl moieties. *Macromol. Chem. Phys.* **2002**, *203*, 1547–1555. [CrossRef]
22. Funahashi, M.; Shimura, H.; Yoshio, M.; Kato, T. Functional Liquid-Crystalline Polymers for Ionic and Electronic Conduction. In *Liquid Crystalline Functional Assemblies and Their Supramolecular Structures (Structure and Bonding)*; Kato, T., Bara, J.E., Eds.; Springer: Berlin, Germany, 2008; Volume 128, pp. 151–179.
23. Germer, R.; Giesselmann, F.; Zugenmaier, P.; Tschierske, C. Anomalous Electric Conductivity in Amphiphilic Smectic Liquid Crystals with Terminal Diol-Groups. *Mol. Cryst. Liq. Cryst.* **1999**, *331*, 643–650. [CrossRef]

24. Yoshio, M.; Mukai, T.; Ohno, H.; Kato, T. One-dimensional ion transport in self-organized columnar ionic liquids. *J. Am. Chem. Soc.* **2004**, *126*, 994–995. [CrossRef]
25. Beginn, U.; Zipp, G.; Möller, M. Functional Membranes Containing Ion-Selective Matrix-Fixed Supramolecular Channels. *Adv. Mater.* **2000**, *12*, 510–513. [CrossRef]
26. Beginn, U.; Zipp, G.; Mourran, A.; Walther, P.; Möller, M. Membranes Containing Oriented Supramolecular Transport Channels. *Adv. Mater.* **2000**, *12*, 513–516. [CrossRef]
27. Charland, J.P.; Buchanan, G.W.; Kirby, R.A. Reinvestigation of the structure of dibenzo-12-crown-4 ether. *Acta Crystallogr. C Cryst. Struct. Commun.* **1989**, *45*, 165–167. [CrossRef]
28. Buchanan, G.W.; Kirby, R.A.; Charland, J.P.; Ratcliffe, C.I. 12-Crown-4 ethers: Solid-state stereochemical features of dibenzo-12-crown-4, derived dicyclohexano-12-crown-4 isomers, and a lithium thiocyanate complex as determined via carbon-13 CPMAS nuclear magnetic resonance and x-ray crystallographic methods. *J. Org. Chem.* **1991**, *56*, 203–212. [CrossRef]
29. Buchanan, G.W.; Mathias, S.; Lear, Y.; Bensimon, C. Dibenzo-15-crown-5 ether and its sodium thiocyanate complex. X-ray crystallographic and NMR studies in the solid phase and in solution. *Can. J. Chem.* **1991**, *69*, 404–414. [CrossRef]
30. Blake, A.J.; Gould, R.O.; Parsons, S.; Radek, C.; Schröder, M. Potassium Dibenzo-18-crown-6 Triiodide. *Acta Crystallogr. C Cryst. Struct. Commun.* **1996**, *52*, 24–27. [CrossRef]
31. Markovitsi, D.; Germain, A.; Millie, P.; Lecuyer, P.; Gallos, L.; Argyrakis, P.; Bengs, H.; Ringsdorf, H. Triphenylene Columnar Liquid Crystals: Excited States and Energy Transfer. *J. Phys. Chem.* **1995**, *99*, 1005–1017. [CrossRef]
32. van Gerwen, P.; Laureyn, W.; Laureys, W.; Huyberechts, G.; de Beeck, M.O.; Baert, K.; Suls, J.; Sansen, W.; Jacobs, P.; Hermans, L.; et al. Nanoscaled interdigitated electrode arrays for biochemical sensors. *Sens. Actuators B* **1998**, *49*, 73–80. [CrossRef]
33. Rodewald, S.; Fleig, J.; Maier, J. Microcontact Impedance Spectroscopy at Single Grain Boundaries in Fe-Doped $SrTiO_3$ Polycrystals. *J. Am. Ceram. Soc.* **2001**, *84*, 521–530. [CrossRef]
34. Stegemeyer, H.; Behret, H. *Liquid Crystals*; Steinkopff: Heidelberg, Germany, 1994; Volume 2.
35. Conejo-Rodriguez, V.; Cuerva, C.; Schmidt, R.; Bardaji, M.; Espinet, P. Li^+ and K^+ ionic conductivity in ionic nematic liquid crystals based on 18-diaza-crown ether substituted with six decylalkoxy-*p*-cyanobiphenyl chains. *J. Mater. Chem. C* **2019**, *7*, 663–672. [CrossRef]

Communication

Substituted Azolium Disposition: Examining the Effects of Alkyl Placement on Thermal Properties

Karel Goossens [1,*,†], Lena Rakers [2,†], Tae Joo Shin [3], Roman Honeker [2], Christopher W. Bielawski [1,4,5,*] and Frank Glorius [2,*]

1   Center for Multidimensional Carbon Materials (CMCM), Institute for Basic Science (IBS), Ulsan 44919, Korea
2   Organisch-Chemisches Institut, Westfälische Wilhelms-Universität Münster, Corrensstraße 40, 48149 Münster, Germany; lena.roling@uni-muenster.de (L.R.); honeker@gmx.de (R.H.)
3   UNIST Central Research Facilities (UCRF) and School of Natural Science, Ulsan National Institute of Science and Technology (UNIST), Ulsan 44919, Korea; tjshin@unist.ac.kr
4   Department of Chemistry, UNIST, Ulsan 44919, Korea
5   Department of Energy Engineering, UNIST, Ulsan 44919, Korea
*   Correspondence: karel.cw.goossens@gmail.com (K.G.); bielawski@unist.ac.kr (C.W.B.); glorius@uni-muenster.de (F.G.); Tel.: +82-52-217-2952 (C.W.B.); +49-251-8333248 (F.G.)
†   These authors contributed equally to this work.

Received: 23 December 2018; Accepted: 6 January 2019; Published: 11 January 2019

Abstract: We describe the thermal phase characteristics of a series of 4,5-bis(*n*-alkyl)azolium salts that were studied using thermogravimetric analysis (TGA), differential scanning calorimetry (DSC), polarized-light optical microscopy (POM), and synchrotron-based small- to wide-angle X-ray scattering (SWAXS) measurements. Key results were obtained for 1,3-dimethyl-4,5-bis(*n*-undecyl)imidazolium iodide (**1-11**), 1,3-dimethyl-4,5-bis(*n*-pentadecyl)imidazolium iodide (**1-15**), and 1,2,3-trimethyl-4,5-bis(*n*-pentadecyl)imidazolium iodide (**2**), which were found to adopt enantiotropic smectic A mesophases. Liquid-crystalline mesophases were not observed for 1,3-dimethyl-4,5-bis(*n*-heptyl)imidazolium iodide (**1-7**), 3-methyl-4,5-bis(*n*-pentadecyl)thiazolium iodide (**3**), and 2-amino-4,5-bis(*n*-pentadecyl)imidazolium chloride (**4**). Installing substituents in the 4- and 5-positions of the imidazolium salts appears to increase melting points while lowering clearing points when compared to data reported for 1,3-disubstituted analogues.

Keywords: liquid crystals; ionic liquid crystals; ionic liquids; imidazolium; thiazolium; mesophases

---

1. Introduction

Attachment of relatively long alkyl or fluoroalkyl chains to the nitrogen atoms of heterocyclic cations, such as imidazolium moieties, typically bestows thermotropic liquid-crystalline (LC) properties due to an enhancement in amphiphilicity [1–4]. Ionic liquid crystals (ILCs) are of high interest because they hold potential for use in a variety of applications [5–9]. For instance, it has already been demonstrated that ILCs may be used as non-volatile electrolytes in dye-sensitized solar cells [10–12], as organized reaction media [13], and as active components in electrochromic devices that do not require additional electrolytes [14–16]. On a more fundamental level, a comprehensive knowledge of the structural parameters that determine the thermal properties of ILCs and the long-range supramolecular organization in their mesophases contributes to a better understanding of the short-range nanosegregation phenomena that occur in room-temperature ionic liquids (ILs) [17–27]. Such structure–property relationships may also improve molecular designs of new ILCs and ILs, and facilitate tailoring for specific applications.

Although imidazolium moieties are among the most widely used cationic cores to obtain ILCs, the majority of studies have focused on *N*-substituted derivatives due to the ease of their synthesis [5–7,28–30].

The influence of installing alkyl or aryl substituents in the 2-position of imidazolium-based ILCs on thermal phase characteristics has also been investigated [31–40]. To the best of our knowledge, only one LC imidazolium salt with a substituent in the 4-position, 1,3,4-trimethylimidazolium *n*-dodecylsulfonate, is known and was reported to adopt a monotropic smectic A (SmA) phase (Cr · (SmA · 93 ·) 95 · Iso (°C)) [32]. For comparison, 1,3-dimethylimidazolium *n*-dodecylsulfonate self-organizes into an enantiotropic SmA phase upon heating and displays a considerably higher clearing point (Cr · 90 · SmA · 177 · Iso (°C)) whereas the corresponding 1,2,3-trimethylimidazolium (Cr · 202 · Iso (°C)) and 1,3,4,5-tetramethylimidazolium (Cr · 109 · Iso (°C)) salts are not LC [32]. Even in the realm of ILs, limited physical property data are available for tri-, tetra-, and pentasubstituted imidazolium salts (with the additional substituents often being limited to methyl groups) [29,30,41–48]. Beyond differences in thermal properties, highly substituted imidazolium salts may be expected to show an increased stability toward base, particularly if they are to be used in, for instance, alkaline fuel cell membranes [49].

Recently, some of us prepared a series of 4,5-bis(*n*-alkyl)imidazolium salts by utilizing Radziszewski-type chemistry for the formation of the requisite heterocyclic cores [50–53]. In addition, 4,5-bis(*n*-pentadecyl)thiazolium salts were synthesized from a thiazole precursor that was obtained by reacting P$_2$S$_5$, formamide and 17-bromodotriacontan-16-one [51]. The bromoketone was also used to prepare a novel guanidinium salt (i.e., 4,5-bis(*n*-pentadecyl)imidazolium with an amino group in the 2-position) [51]. The aforementioned salts were studied as lipid analogues in model cell systems or used as precursors for *N*-heterocyclic carbenes for the stabilization of nanoparticles [50–53]. From the library of compounds, we selected six 4,5-bis(*n*-alkyl)azolium salts (**1-*n*** (*n* = 7, 11, 15) and **2–4**, Scheme 1) and investigated their thermal characteristics. The data were compared with those reported for analogous 1,3-bis(*n*-alkyl)imidazolium and 2-[3,4-bis(*n*-alkyloxy)phenyl]-1,3- dimethylimidazolium ILCs (for example, **5-*n*** and **6-*n*** in Scheme 1) [38,54–59]. Collectively, the results that will be described below show how the disposition of long alkyl chains around five-membered, heteroaromatic, cationic cores affects the thermal characteristics of the resulting salts.

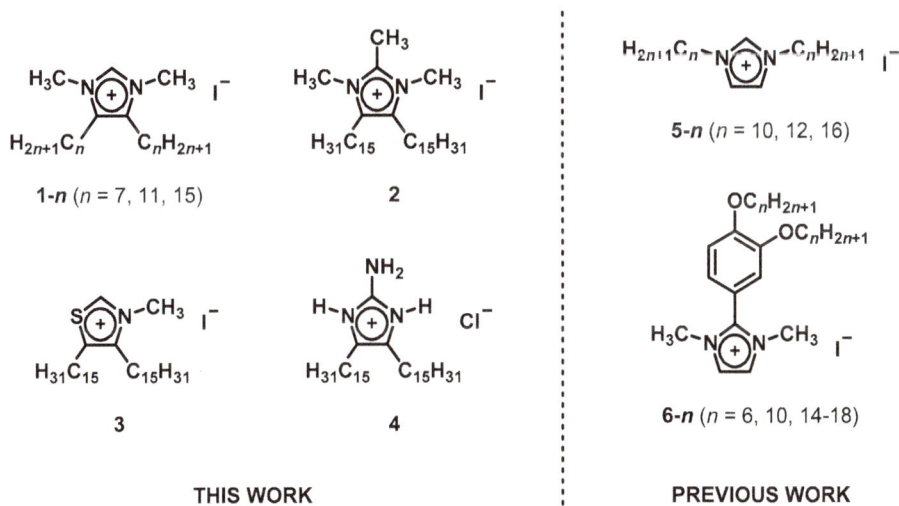

Scheme 1. Left: overview of the 4,5-bis(*n*-alkyl)azolium salts that were investigated in this work. Right: previously reported imidazolium salts that feature two relatively long alkyl chains [38,54].

2. Materials and Methods

The synthesis of compounds **1-*n*** and **2–4** has been described elsewhere [50,51].

Optical textures were observed using an Olympus BX53-P polarized-light optical microscope that was equipped with a rotatable graduated sample platform and an Instec HCS402 dual heater temperature stage. The latter was equipped with a precision XY positioner, and was coupled to an Instec LN$_2$-SYS liquid nitrogen cooling system and an Instec mK2000 programmable temperature controller. Images were recorded by a QImaging Retiga 2000R CCD camera that was coupled to the microscope. The samples were pressed between an untreated glass slide and glass coverslip (0.13–0.17 mm thick, Duran) prior to analysis.

Differential scanning calorimetry (DSC) data were recorded under nitrogen (50 mL·min^{-1}) on a TA Instruments DSC Q2000 module equipped with an RCS90 cooling system at a heating rate of 10 °C·min^{-1} and a cooling rate of 5 °C·min^{-1}. The quantity of sample analyzed was typically 4–5 mg. A small hole was pierced into the lid of the aluminum sample pans. The measurements were performed using the Tzero^{TM} Heat Flow T4P option. High-purity sapphire disks were used for the Tzero^{TM} calibration and high-purity indium was used as a standard for temperature and enthalpy calibrations. DSC data analysis was performed with the Universal Analysis 2000 software (version 4.5A) from TA Instruments (New Castle, DE, USA). The abbreviations used to describe the thermal phase properties are explained in the main text and in the captions of the figures and tables.

Thermogravimetric analysis (TGA) data were recorded under nitrogen (60 mL·min^{-1}) on a TA Instruments TGA Q500 module at a heating rate of 5 °C·min^{-1} and using a platinum sample pan. The quantity of sample analyzed was typically 5–8 mg. High-purity nickel was used as a standard for temperature calibration (based on its Curie temperature).

Synchrotron-based X-ray scattering measurements were performed at the PLS-II 6D UNIST-PAL Beamline of the Pohang Accelerator Laboratory (PAL), Pohang, Republic of Korea. The X-rays coming from the bending magnet were monochromated using Si(111) double crystals and focused at the detector position by the combination of a second, sagittal-type monochromator crystal and a toroidal mirror system. The diffraction patterns were recorded by a Rayonix MX225-HS 2D CCD detector (225 × 225 mm^2 square active area, full resolution 5760 × 5760 pixels) with 2 × 2 binning. The peak positions in the 1D intensity profiles, which were obtained from azimuthal averaging of the 2D patterns of non-aligned samples (with dezingering applied to the data of two separate measurements), were used for phase type assignments. Small- to wide-angle X-ray scattering (SWAXS) patterns (for periodicities up to 67 Å) were recorded using 12.3984 keV X-ray radiation (wavelength λ = 1.00 Å) and a sample-to-detector distance (SDD) of ca. 431 mm. Diffraction angles were calibrated using a lanthanum hexaboride (LaB$_6$) standard (NIST SRM 660c). Samples were contained in borosilicate glass (glass #50) capillaries with an outer diameter of 0.4 mm and a wall thickness of 10 μm and were irradiated for 10–30 s per measurement, depending on the saturation level of the detector. The capillaries were inserted into a custom-made brass holder that was placed into a Linkam TS1500V heating stage to achieve temperature control. The samples were allowed to equilibrate at each temperature before starting a measurement.

Molecular models were created using the Chem3D Pro 15.1 software package (PerkinElmer Informatics, Inc., Waltham, MA, USA).

3. Results

We examined the thermal properties of 4,5-bis(*n*-alkyl)azolium salts **1-*n*** and **2-4** using TGA, DSC, and polarized-light optical microscopy (POM). Key results are summarized in Table 1. Figures showing the TGA and DSC thermograms can be found in the Supplementary Material (Figures S1–S7).

Table 1. Summary of phase transition temperatures and other thermal data recorded for the 4,5-bis(*n*-alkyl)azolium salts **1-*n*** (*n* = 7, 11, 15) and **2-4**, as well as corresponding assignments.

Compound	Transition [1]		T (°C) [2]	ΔH (kJ·mol^{-1}) [3]	$T_{1\%}$ (°C) [4]
1-7		Cr → Iso [5]	26	0.7	n.d.
1-11	*HR1*:	Cr → SmA [6]	50, 55 [6]	21.6 [6]	n.d.
		SmA → Iso	77	2.3	
	HR2:	g → SmA	~−14	−	
		SmA → Iso	76	2.2	
1-15		Cr → SmA	76	38.3	~181
		SmA → Iso	88 [7]	0.7	
2		Cr → SmA [8]	70 [8]	− [8]	~194
		SmA → Iso	88 [7]	0.5	
3		Cr → Iso [9]	73 [7]	16.4	~135
4		Cr → Iso [9]	68 [7]	38.1	~206

[1] Abbreviations: Cr = crystalline phase; g = glass; SmA = smectic A phase; Iso = isotropic liquid phase. HR1 = first heating run; HR2 = second heating run. [2] Onset temperatures obtained by DSC during the second heating run (unless indicated otherwise) at a rate of 10 °C·min^{-1} and under an atmosphere of N$_2$ (50 mL·min^{-1}). A small hole was pierced into the lid of the DSC sample pans. The glass transition temperature was defined by the inflection point of the signal recorded in the respective DSC thermogram. [3] Enthalpy change. [4] Temperature at which 1% weight loss was measured by TGA (neglecting initial small weight losses attributed to the release of H$_2$O). n.d. = not determined. [5] During the second heating run, melting was preceded by an exothermic recrystallization event with a peak temperature of 22 °C (ΔH = −0.4 kJ·mol^{-1}) (Figure S1). [6] This transition involved two, partially resolved transitions between 37 °C and 68 °C (Figure S2). The peak temperatures of the two signals and the total enthalpy change are listed. Examination by POM revealed that the sample became plastic after the first transition upon heating. [7] Peak temperature. [8] The transition to the LC mesophase upon heating was preceded by multiple transitions (Figure S4). The peak temperature of the last endothermic signal just before the temperature range of the SmA phase is given. Examination by POM revealed that the sample gradually became plastic above ~50 °C upon heating. The enthalpy change associated with the transition from the SmA phase to the solid state at 40 °C upon cooling was measured to be 12.2 kJ·mol^{-1}. [9] Melting was preceded by solid-to-solid transitions (Figures S5 and S6).

The slightly higher thermal stability measured for **2** as compared to salt **1-15** can be ascribed to the replacement of the acidic H(2) proton of the imidazolium with a methyl group. The result is in accordance with the generally observed higher thermal stabilities of 2-methylsubstituted 1,3-dialkylimidazolium ILs relative to their unsubstituted counterparts [60]. The 3,4,5-trialkylsubstituted thiazolium salt **3** showed a relatively low thermal stability, with weight losses starting to occur at about 135 °C under the conditions employed for the TGA measurements. The most thermally stable salt among the compounds that were studied was **4** ($T_{1\%}$ ≈ 206 °C), despite the fact that it contains nucleophilic Cl$^-$ anions as well as amino groups.

DSC and POM investigations of compounds **1-*n*** and **2–4** led to the conclusion that imidazoliums **1-11**, **1-15**, and **2** are thermotropic ILCs but imidazolium **1-7**, thiazolium **3**, and guanidinium **4** are not. To the best of our knowledge, **2** is the first example of a LC pentasubstituted imidazolium salt. Salt **1-7**, which was obtained as a partially crystallized compound, melted to an isotropic liquid around room temperature. In contrast, solid samples of **1-11**, **1-15**, and **2**, which are more amphiphilic than **1-7**, melted to birefringent mesophases which produced "oily streaks" when viewed by POM (Figure 1). Homeotropic domains were observed as well. Further heating facilitated transitions to isotropic liquid states, for which, in the case of **1-15** and **2**, only weak signals were observed in the DSC thermograms (Figure 2 and Figures S2–S4). Upon cooling, ill-shaped "bâtonnets" were seen by POM upon entering the LC states and gradually transformed into focal-conic-like textures with further cooling. In the case of **1-11**, the sample spontaneously aligned homeotropically when cooled from the isotropic liquid state and the LC mesophase vitrified around −23 °C; glass transitions continued to be seen during subsequent heating/cooling cycles (Figure 2a). The non-mesomorphic salts **3** and **4** melted directly to isotropic liquids without passing through intermediate LC phases. The melting point data recorded for all 4,5-bis(*n*-pentadecyl)azolium salts (**1-15** and **2–4**) are similar and situated in the range of 68 to 76 °C.

(a)

(b)

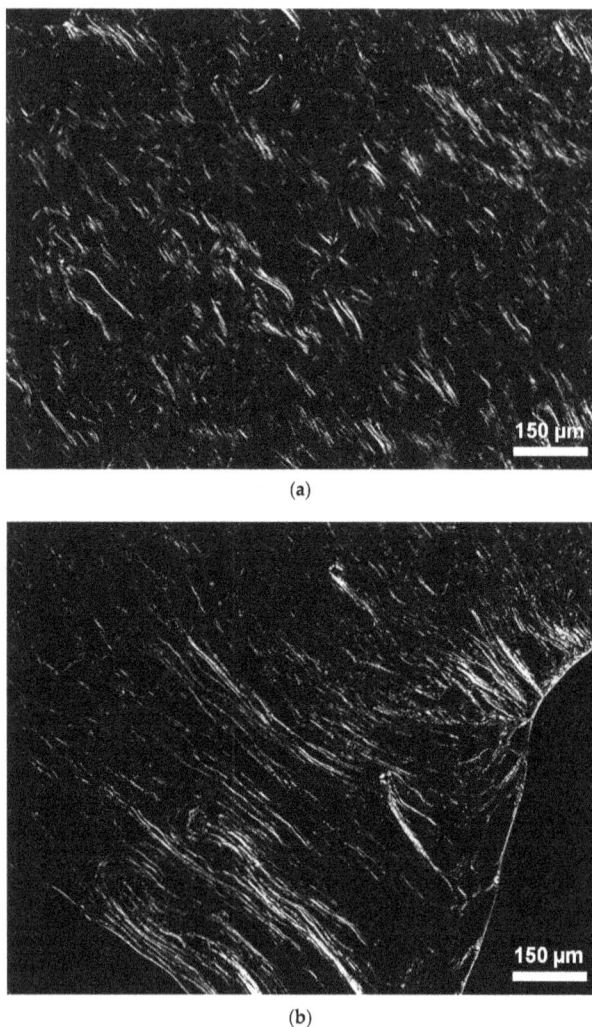

Figure 1. Polarized-light optical microscopy (POM) images of the SmA phases of (a) **1-11** at 24 °C (upon cooling from the isotropic liquid state and after applying pressure to the sample) and (b) **2** at 77 °C (during the first heating run of a pristine sample).

Synchrotron-based small- to wide-angle X-ray scattering (SWAXS) measurements of the LC mesophases adopted by **1-15** and **2** afforded patterns that were characterized by two sharp reflections in the small-angle region, in addition to a diffuse wide-angle scattering signal centered at 4.4–4.6 Å (Figure 3, Table 2 and Figure S8). The latter corresponds to the lateral short-range order of the molten alkyl chains (c.f. h_{ch}) and the ionic headgroups (c.f. h_{ion}), respectively, which were not resolved in the experimental data. The reciprocal *d*-spacings of the sharp small-angle reflections were related by a 1:2 ratio and the signals can be indexed as the (001) and (002) reflections that originate from the formation of layers. Collectively, the POM and SWAXS data indicate that the LC mesophases adopted by salts **1-15** and **2** are SmA phases. A slightly larger layer thickness was found for **2** as compared with **1-15** at similar temperatures (see Table 2), which can be ascribed to the protruding 2-methyl substituents in the former. Based on the POM observations, the mesophase adopted by **1-11** is also a SmA phase.

SWAXS patterns that were recorded for **3** at different temperatures revealed that the sample adopted a lamellar structure in the solid state before melting to an isotropic liquid at about 73 °C (Figure S9).

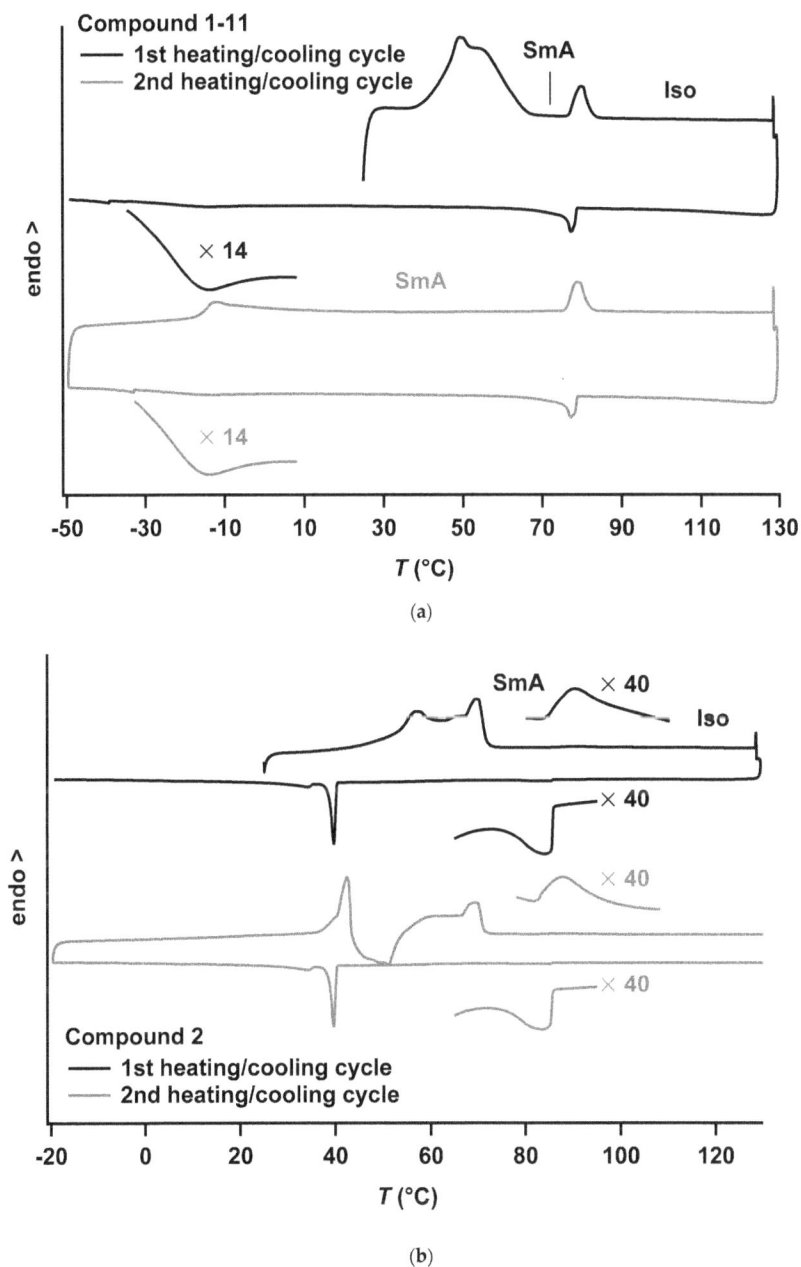

Figure 2. Differential scanning calorimetry (DSC) data recorded for (**a**) **1-11** and (**b**) **2** at a heating rate of 10 °C·min^{-1} and a cooling rate of 5 °C min^{-1} under an atmosphere of N$_2$. Endothermic peaks point upward.

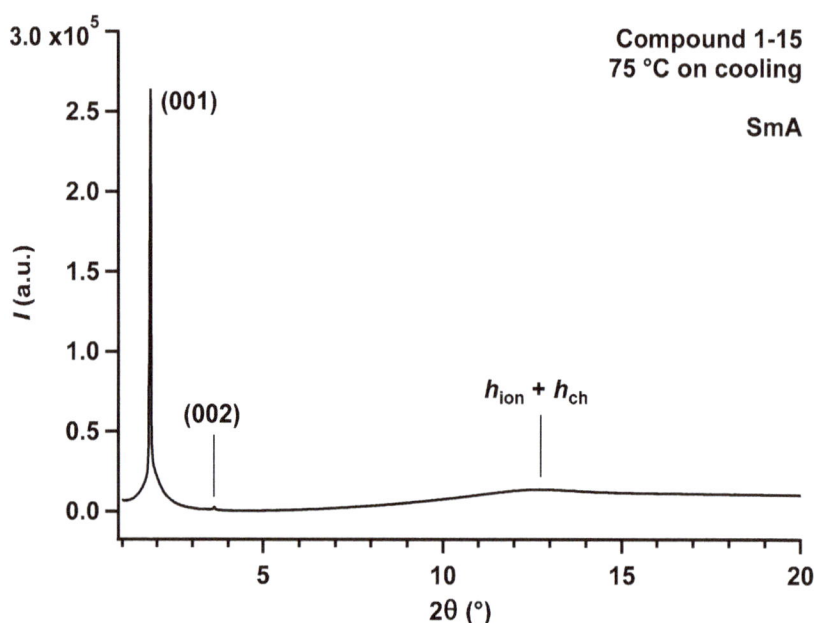

Figure 3. Synchrotron-based SWAXS data that were recorded for the SmA phase of **1-15** at 75 °C upon cooling (the X-ray wavelength used was 1.00 Å).

Table 2. Summary of synchrotron-based SWAXS data recorded for the LC mesophases adopted by **1-15** and **2**, including calculated structural parameters and corresponding assignments.

Cpd.	Type of LC Mesophase	T (°C)	$d_{obs.}$ (Å) [1]	I [2]	hkl [3]	$d_{calcd.}$ (Å) [1]	Structural Parameters of the LC Mesophases [4]
1-15	SmA	75	31.63	VS (sh)	001	31.68	d = 31.68 Å
	(upon		15.86	W (sh)	002	15.84	V_{mol} ≈ 1072 Å^3
	cooling)		4.5–4.6	br	h_1		A_M ≈ 67.7 Å^2
							σ_{ch} ≈ 22.1 Å^2
2	SmA	78	32.91	VS (sh)	001	32.88	d = 32.88 Å
			16.42	W (sh)	002	16.44	V_{mol} ≈ 1098 Å^3
			4.4–4.5	br	h_1		A_M ≈ 66.8 Å^2
							σ_{ch} ≈ 22.2 Å^2

[1] The $d_{obs.}$ and $d_{calcd.}$ values refer to the measured and calculated diffraction spacings, respectively. $d_{calcd.} = <d_{001}> = [\Sigma_l d_{00l} l]/N_{00l}$, in which N_{00l} = the number of (00l) reflections. [2] I is the intensity of each reflection: VS = very strong, W = weak, sh = sharp reflection, and br = broad reflection. [3] hkl are the Miller indices of the reflections. h_1 indicates the center position of the diffuse wide-angle signal that originates from the lateral short-range order of the ionic moieties (c.f. h_{ion}) and the molten alkyl chains (c.f. h_{ch}). [4] V_{mol} is the molecular volume, which was estimated as $V_{mol}(T) = (M_{cation}/0.6022)f + V_{iodide}$, in which M_{cation} is the molecular mass of the cations (in g·mol^{-1}), f is a temperature-correcting factor (f = 0.9813 + 7.474 × 10^{-4}T with T in °C) and V_{iodide} is the partial volume of the iodide anions as determined from reference salts [61]. A_M is the cross-sectional area that is occupied by molecular assemblies along the sequence of smectic layers and was calculated as $A_M(T) = 2V_{mol}(T)/d(T)$ [38,61]. The cross-sectional area of one fully stretched aliphatic chain, σ_{ch} [62], is listed for comparison.

From the structural parameters that were obtained from the SWAXS measurements as well as the temperature-dependent molecular volumes calculated for **1-15** and **2**, it can be concluded that the SmA phases are characterized by alternating, nanosegregated ionic and aliphatic sublayers, with a head-to-head arrangement of the ionic headgroups in the former, and partially interdigitated and folded alkyl chains in the latter [63,64]. The values calculated for the molecular cross-sectional areas, A_M (see Table 2), are comparable to those previously reported for the SmA phases adopted by

2-aryl-1,3-dimethylimidazolium iodide salts having two alkyl chains per cation (**6-*n*** (*n* = 6, 10, 14)) [38]. The supramolecular arrangements found in all of these SmA phases are similar. We note that **1-15** has been reported to spontaneously form thermodynamically stable vesicles in buffered aqueous media without the addition of other lipids [52]. As such, the structure of its thermotropic LC mesophase may resemble the local structure of the vesicle bilayer membranes. It was also found that **1-7** does not form bilayer vesicles under similar conditions [52]. Herein we report that it does not form a thermotropic LC mesophase either.

4. Discussion

The observation of SmA mesophases for imidazolium salts **1-*n*** (*n* = 11, 15) and **2** was not unexpected, since structurally related 1,3-bis(*n*-alkyl)imidazolium salts with long alkyl chains are known to adopt thermotropic SmA phases [54–59]. However, inspection of the literature data, some of which are collected in Table 3, shows that the clearing points of the 4,5-disubstituted ILCs are lower than those of 1,3-analogues. For example, the alkyl substituents of iodide salt **5-16** contain only one more methylene group than **1-15** or **2**, yet the clearing point of the former is 59 °C higher and its SmA phase is stable over a temperature range of 80 °C [54]. Even the homologue with *n*-dodecyl groups in the 1- and 3-positions, **5-12**, exhibits a similar clearing point as **1-15** and **2** despite its lower amphiphilic character. Since **5-12** also has a lower melting point, its mesophase temperature range is about 49 °C [56]. Likewise, [C$_{16}$C$_{16}$im][BF$_4$] and [C$_{16}$C$_{16}$im][PF$_6$] display higher clearing points and, thus, more stable SmA phases than **1-15** or **2** (we note that [BF$_4$]$^-$ has a similar volume as the iodide anion, while [PF$_6$]$^-$ is about 1.5 times as large [65]) [54,57]. Compound **1-11** also has a higher melting point and lower clearing point than **5-12** [56]. As such, 1,3-disubstitution appears to induce thermotropic LC mesomorphism in imidazolium salts more effectively than the 4,5-disubstituted analogues.

Table 3. A comparison of the thermal phase characteristics exhibited by 1,3-bis(*n*-alkyl)substituted imidazolium salts and the 4,5-bis(*n*-alkyl)imidazolium salts **1-11**, **1-15**, and **2**.

Compound [1]	Phase Transition Temperatures (°C) [2]
1-7	Cr · 26 · Iso
1-11	HR1: Cr · ~55 [3] · SmA · 77 · Iso
	HR2: g · ~−14 · SmA · 76 · Iso
1-15	Cr · 76 · SmA · 88 · Iso
2	Cr · 70 · SmA · 88 · Iso
[C$_{10}$C$_{10}$im][I] (**5-10**) [54]	Cr · <0 · SmA · 55 · Iso
[C$_{12}$C$_{12}$im][I] (**5-12**) [56]	Cr · 40 · SmA · 89 · Iso
[C$_{16}$C$_{16}$im][I] (**5-16**) [54]	Cr · 67 · SmA · 147 · Iso
[C$_{10}$C$_{10}$im][BF$_4$] [57]	Cr · 18 · SmA · 25 · Iso
[C$_{12}$C$_{12}$im][BF$_4$] [57]	Cr · 50 · SmA · 69 · Iso
[C$_{14}$C$_{14}$im][BF$_4$] [57]	Cr · 63 · SmA · 106 · Iso
[C$_{16}$C$_{16}$im][BF$_4$] [57]	Cr · 70 · SmA · 125 · Iso
[C$_{10}$C$_{10}$im][PF$_6$] [54,66]	Cr · 16 · Iso
[C$_{12}$C$_{12}$im][PF$_6$] [54,57]	Cr · 45 · Iso
[C$_{14}$C$_{14}$im][PF$_6$] [54]	Cr · 59 · SmA · 81 · Iso
[C$_{16}$C$_{16}$im][PF$_6$] [54]	Cr · 68 · SmA · 105 · Iso

[1] [C$_n$C$_n$im]$^+$ = 1,3-bis(*n*-alkyl)imidazolium, where the subscript *n* indicates the number of carbon atoms in the alkyl chains. [2] Abbreviations: Cr = crystalline phase; g = glass; SmA = smectic A phase; Iso = isotropic liquid phase. HR1 = first heating run; HR2 = second heating run. [3] See Table 1.

While the reasons underlying the aforementioned deduction are unclear, the imidazolium H(4) and H(5) atoms are known to participate in hydrogen bonds with halide counterions (see, for example, reference [67]), which may stabilize the ionic sublayers in the smectic mesophases. Replacement of the hydrogen atoms by long alkyl chains may impede such hydrogen-bond interactions. The steric parameters of the methyl groups in the 1- and 3-positions may also hinder compact arrangement of the ionic headgroups in the ionic sublayers, although: (1) similar A_M values were found for the SmA phases adopted by 2-aryl-1,3-dimethylimidazolium iodide salts having two alkyl chains per cation (see above) and those smectic phases were stable until 147–164 °C [38], and (2) LC mesophases were not detected for thiazolium salt **3** which features only one *N*-methyl group. The latter observation provides another example of how subtle structural and electronic changes in organic salts may have a considerable impact on their physical properties. Although LC thiazolium salts have not yet been reported to the best of our knowledge, the apparent absence of mesomorphic properties for compound **3** does not exclude the possibility that 3-alkylthiazolium salts with relatively long alkyl chains may show such characteristics.

We tentatively ascribe the higher melting point of imidazoliums **1-11** and **1-15** as compared with **5-12** and **5-16**, respectively, to a closer proximity of the alkyl chains in the former. Such arrangement may facilitate van der Waals interactions between the chains and increase the melting point.

As mentioned above, a LC phase was not observed for **4**. The potential beneficial effect of the N–H sites in the cationic headgroup, which could participate in hydrogen bonds within ionic sublayers, may be counteracted by a mismatch in cross-sectional area between the ionic moieties and the molten, long alkyl chains, which is a prerequisite for the development of a smectic LC mesophase. The latter effect may explain the direct transition to an isotropic liquid and the lowest measured melting point among the series (**1-15**)–**4**.

We also note that **6-15** exhibits enantiotropic cubic and columnar LC mesophases while homologues with shorter alkyl chains adopt only SmA phases [38]. The absence of non-smectic LC phases for the salts discussed herein underscores the importance of the "taper angles" and precise geometric shapes of the polar headgroups of amphiphilic mesogens for inducing columnar and bicontinuous cubic mesophases [68–70].

5. Conclusions

We report the thermal properties for a series of 4,5-bis(*n*-alkyl)azolium salts. The majority of previously reported imidazolium- and thiazolium-based ionic liquids and ionic liquid crystals featured 1,3-disubstituted or *N*-substituted cations, respectively. The attachment of two long alkyl chains in the 4- and 5-positions of imidazolium or thiazolium rings is unusual because it requires a different synthetic approach than simply alkylating commercially available precursors, such as 1-methylimidazole, imidazole, or thiazole. While 1,3-dimethyl-4,5-bis(*n*-heptyl)imidazolium iodide (**1-7**) is non-mesomorphic, 1,3-dimethyl-4,5-bis(*n*-undecyl)imidazolium iodide (**1-11**), 1,3-dimethyl-4,5-bis(*n*-pentadecyl)imidazolium iodide (**1-15**) and the analogue with an additional methyl group in the 2-position (**2**) were found to be liquid-crystalline; they adopt SmA phases upon heating albeit over relatively narrow temperature ranges. For comparison, 3-methyl-4,5-bis(*n*-pentadecyl)thiazolium iodide (**3**) or 2-amino-4,5-bis(*n*-pentadecyl)imidazolium chloride (**4**) did not display liquid-crystalline properties under the conditions explored. Regardless, installing relatively long alkyl substituents in the 4- and 5-positions of the imidazolium salts appears to result in increased melting points and lowered clearing points, particularly when compared to data reported for 1,3-disubstituted analogues. The observation may be attributed to changes in hydrogen-bonding interactions, increased steric requirements of the ionic headgroups due to the *N*-methyl groups, and/or differences in proximity of the long alkyl chains. It remains to be seen whether non-smectic mesophases can be formed by 4,5-dialkylazolium salts. Further synthetic developments may be needed to answer these questions.

Supplementary Materials: The following are available online at http://www.mdpi.com/2073-4352/9/1/34/s1, Figures S1–S6: DSC data recorded for compounds **1-*n*** (*n* = 7, 11, 15) and **2–4**, Figure S7: TGA data recorded for compounds **1-15** and **2–4**, Figure S8: synchrotron-based SWAXS pattern that was recorded for compound **2** at 78 °C, Figure S9: synchrotron-based SWAXS patterns that were recorded for compound **3** at different temperatures.

Author Contributions: K.G. and F.G. conceived the project; C.W.B. supervised the project; L.R. synthesized compounds **1-*n*** (*n* = 7, 11, 15), **2**, and **3**; R.H. synthesized compound **4**; K.G. performed the thermal and structural characterizations (TGA, DSC, POM, synchrotron-based SWAXS) and analyzed the data; T.J.S. supervised the synchrotron-based SWAXS measurements; K.G. wrote the manuscript with contributions from all authors.

Funding: K.G. and C.W.B. were supported in part by the Institute for Basic Science (IBS-R019-D1). C.W.B. acknowledges the BK21 Plus Program as funded by the Ministry of Education and the National Research Foundation of Korea for support. L.R., R.H. and F.G. gratefully acknowledge the Deutsche Forschungsgemeinschaft (SFB 858). The experiments at the PLS-II 6D UNIST-PAL Synchrotron Beamline were supported in part by the Korean Ministry of Science and ICT (MSIT), Pohang University of Science and Technology (POSTECH) and the UNIST Central Research Facilities (UCRF).

Acknowledgments: We thank Seong-Hun Lee and Juhyun Yang for their assistance with the synchrotron measurements.

Conflicts of Interest: The authors declare no conflict of interest. The funders had no role in the design of the study; in the collection, analyses, or interpretation of data; in the writing of the manuscript, or in the decision to publish the results.

References

1. Bowlas, C.J.; Bruce, D.W.; Seddon, K.R. Liquid-Crystalline Ionic Liquids. *Chem. Commun.* **1996**, 1625–1626. [CrossRef]

2. Gordon, C.M.; Holbrey, J.D.; Kennedy, A.R.; Seddon, K.R. Ionic Liquid Crystals: Hexafluorophosphate Salts. *J. Mater. Chem.* **1998**, *8*, 2627–2636. [CrossRef]

3. Holbrey, J.D.; Seddon, K.R. The Phase Behaviour of 1-Alkyl-3-Methylimidazolium Tetrafluoroborates; Ionic Liquids and Ionic Liquid Crystals. *J. Chem. Soc. Dalton Trans.* **1999**, 2133–2139. [CrossRef]

4. Goossens, K.; Lava, K.; Nockemann, P.; Van Hecke, K.; Van Meervelt, L.; Driesen, K.; Görller-Walrand, C.; Binnemans, K.; Cardinaels, T. Pyrrolidinium Ionic Liquid Crystals. *Chem. Eur. J.* **2009**, *15*, 656–674. [CrossRef] [PubMed]

5. Goossens, K.; Lava, K.; Bielawski, C.W.; Binnemans, K. Ionic Liquid Crystals: Versatile Materials. *Chem. Rev.* **2016**, *116*, 4643–4807. [CrossRef] [PubMed]

6. Mansueto, M.; Laschat, S. Ionic Liquid Crystals. In *Handbook of Liquid Crystals. Volume 6: Nanostructured and Amphiphilic Liquid Crystals*, 2nd ed.; Goodby, J.W., Collings, P.J., Kato, T., Tschierske, C., Gleeson, H., Raynes, P., Eds.; Wiley-VCH: Weinheim, Germany, 2014; pp. 231–280.

7. Fernandez, A.A.; Kouwer, P.H.J. Key Developments in Ionic Liquid Crystals. *Int. J. Mol. Sci.* **2016**, *17*, 731. [CrossRef] [PubMed]

8. Kato, T.; Yoshio, M.; Ichikawa, T.; Soberats, B.; Ohno, H.; Funahashi, M. Transport of Ions and Electrons in Nanostructured Liquid Crystals. *Nat. Rev. Mater.* **2017**, *2*, 17001. [CrossRef]

9. Kato, T.; Uchida, J.; Ichikawa, T.; Sakamoto, T. Functional Liquid Crystals towards the Next Generation of Materials. *Angew. Chem. Int. Ed.* **2018**, *57*, 4355–4371. [CrossRef]

10. Yamanaka, N.; Kawano, R.; Kubo, W.; Kitamura, T.; Wada, Y.; Watanabe, M.; Yanagida, S. Ionic Liquid Crystal as a Hole Transport Layer of Dye-Sensitized Solar Cells. *Chem. Commun.* **2005**, 740–742. [CrossRef]

11. Yamanaka, N.; Kawano, R.; Kubo, W.; Masaki, N.; Kitamura, T.; Wada, Y.; Watanabe, M.; Yanagida, S. Dye-Sensitized TiO_2 Solar Cells Using Imidazolium-Type Ionic Liquid Crystal Systems As Effective Electrolytes. *J. Phys. Chem. B* **2007**, *111*, 4763–4769. [CrossRef]

12. Högberg, D.; Soberats, B.; Yatagai, R.; Uchida, S.; Yoshio, M.; Kloo, L.; Segawa, H.; Kato, T. Liquid-Crystalline Dye-Sensitized Solar Cells: Design of Two-Dimensional Molecular Assemblies for Efficient Ion Transport and Thermal Stability. *Chem. Mater.* **2016**, *28*, 6493–6500. [CrossRef]

13. Bruce, D.W.; Gao, Y.A.; Canongia Lopes, J.N.; Shimizu, K.; Slattery, J.M. Liquid-Crystalline Ionic Liquids As Ordered Reaction Media for the Diels-Alder Reaction. *Chem. Eur. J.* **2016**, *22*, 16113–16123. [CrossRef] [PubMed]

14. Yazaki, S.; Funahashi, M.; Kato, T. An Electrochromic Nanostructured Liquid Crystal Consisting of π-Conjugated and Ionic Moieties. *J. Am. Chem. Soc.* **2008**, *130*, 13206–13207. [CrossRef] [PubMed]

15. Yazaki, S.; Funahashi, M.; Kagimoto, J.; Ohno, H.; Kato, T. Nanostructured Liquid Crystals Combining Ionic and Electronic Functions. *J. Am. Chem. Soc.* **2010**, *132*, 7702–7708. [CrossRef] [PubMed]

16. Beneduci, A.; Cospito, S.; La Deda, M.; Veltri, L.; Chidichimo, G. Electrofluorochromism in π-Conjugated Ionic Liquid Crystals. *Nat. Commun.* **2014**, *5*, 3105. [CrossRef] [PubMed]

17. Hayes, R.; Warr, G.G.; Atkin, R. Structure and Nanostructure in Ionic Liquids. *Chem. Rev.* **2015**, *115*, 6357–6426. [CrossRef] [PubMed]

18. Ji, Y.; Shi, R.; Wang, Y.; Saielli, G. Effect of the Chain Length on the Structure of Ionic Liquids: From Spatial Heterogeneity to Ionic Liquid Crystals. *J. Phys. Chem. B* **2013**, *117*, 1104–1109. [CrossRef]
19. Nemoto, F.; Kofu, M.; Yamamuro, O. Thermal and Structural Studies of Imidazolium-Based Ionic Liquids with and Without Liquid-Crystalline Phases: The Origin of Nanostructure. *J. Phys. Chem. B* **2015**, *119*, 5028–5034. [CrossRef]
20. Russina, O.; Lo Celso, F.; Plechkova, N.; Jafta, C.J.; Appetecchi, G.B.; Triolo, A. Mesoscopic Organization in Ionic Liquids. *Top. Curr. Chem.* **2017**, *375*, 58. [CrossRef]
21. Russina, O.; Lo Celso, F.; Plechkova, N.V.; Triolo, A. Emerging Evidences of Mesoscopic-Scale Complexity in Neat Ionic Liquids and Their Mixtures. *J. Phys. Chem. Lett.* **2017**, *8*, 1197–1204. [CrossRef]
22. Bruce, D.W.; Cabry, C.P.; Canongia Lopes, J.N.; Costen, M.L.; D'Andrea, L.; Grillo, I.; Marshall, B.C.; McKendrick, K.G.; Minton, T.K.; Purcell, S.M.; et al. Nanosegregation and Structuring in the Bulk and at the Surface of Ionic-Liquid Mixtures. *J. Phys. Chem. B* **2017**, *121*, 6002–6020. [CrossRef] [PubMed]
23. Pontoni, D.; Haddad, J.; Di Michiel, M.; Deutsch, M. Self-Segregated Nanostructure in Room Temperature Ionic Liquids. *Soft Matter* **2017**, *13*, 6947–6955. [CrossRef]
24. Cosby, T.; Vicars, Z.; Wang, Y.; Sangoro, J. Dynamic-Mechanical and Dielectric Evidence of Long-Lived Mesoscale Organization in Ionic Liquids. *J. Phys. Chem. Lett.* **2017**, *8*, 3544–3548. [CrossRef] [PubMed]
25. Nemoto, F.; Kofu, M.; Nagao, M.; Ohishi, K.; Takata, S.; Suzuki, J.; Yamada, T.; Shibata, K.; Ueki, T.; Kitazawa, Y.; et al. Neutron Scattering Studies on Short- and Long-Range Layer Structures and Related Dynamics in Imidazolium-Based Ionic Liquids. *J. Chem. Phys.* **2018**, *149*, 054502. [CrossRef] [PubMed]
26. Cabry, C.P.; D'Andrea, L.; Shimizu, K.; Grillo, I.; Li, P.; Rogers, S.; Bruce, D.W.; Canongia Lopes, J.N.; Slattery, J.M. Exploring the Bulk-Phase Structure of Ionic Liquid Mixtures Using Small-Angle Neutron Scattering. *Faraday Discuss.* **2018**, *206*, 265–289. [CrossRef] [PubMed]
27. Cao, W.D.; Wang, Y.T.; Saielli, G. Metastable State During Melting and Solid-Solid Phase Transition of $[C_n Mim][NO_3]$ (n = 4-12) Ionic Liquids by Molecular Dynamics Simulation. *J. Phys. Chem. B* **2018**, *122*, 229–239. [CrossRef] [PubMed]
28. Douce, L.; Suisse, J.-M.; Guillon, D.; Taubert, A. Imidazolium-Based Liquid Crystals: A Modular Platform for Versatile New Materials with Finely Tuneable Properties and Behaviour. *Liq. Cryst.* **2011**, *38*, 1653–1661. [CrossRef]
29. Bara, J.E.; Shannon, M.S. Beyond 1,3-Difunctionalized Imidazolium Cations. *Nanomater. Energy* **2012**, *1*, 237–242. [CrossRef]
30. Scalfani, V.F.; Alshaikh, A.A.; Bara, J.E. Analysis of the Frequency and Diversity of 1,3-Dialkylimidazolium Ionic Liquids Appearing in the Literature. *Ind. Eng. Chem. Res.* **2018**, *57*, 15971–15981. [CrossRef]
31. Fox, D.M.; Awad, W.H.; Gilman, J.W.; Maupin, P.H.; De Long, H.C.; Trulove, P.C. Flammability, Thermal Stability, and Phase Change Characteristics of Several Trialkylimidazolium Salts. *Green Chem.* **2003**, *5*, 724–727. [CrossRef]
32. Mukai, T.; Yoshio, M.; Kato, T.; Ohno, H. Effect of Methyl Groups Onto Imidazolium Cation Ring on Liquid Crystallinity and Ionic Conductivity of Amphiphilic Ionic Liquids. *Chem. Lett.* **2004**, *33*, 1630–1631. [CrossRef]
33. Mukai, T.; Yoshio, M.; Kato, T.; Yoshizawa-Fujita, M.; Ohno, H. Self-Organization of Protonated 2-Heptadecylimidazole as an Effective Ion Conductive Matrix. *Electrochemistry* **2005**, *73*, 623–626.
34. Kouwer, P.H.J.; Swager, T.M. Synthesis and Mesomorphic Properties of Rigid-Core Ionic Liquid Crystals. *J. Am. Chem. Soc.* **2007**, *129*, 14042–14052. [CrossRef] [PubMed]
35. Yoshio, M.; Ichikawa, T.; Shimura, H.; Kagata, T.; Hamasaki, A.; Mukai, T.; Ohno, H.; Kato, T. Columnar Liquid-Crystalline Imidazolium Salts. Effects of Anions and Cations on Mesomorphic Properties and Ionic Conductivity. *Bull. Chem. Soc. Jpn.* **2007**, *80*, 1836–1841. [CrossRef]
36. Alam, M.A.; Motoyanagi, J.; Yamamoto, Y.; Fukushima, T.; Kim, J.; Kato, K.; Takata, M.; Saeki, A.; Seki, S.; Tagawa, S.; et al. "Bicontinuous Cubic" Liquid Crystalline Materials from Discotic Molecules: A Special Effect of Paraffinic Side Chains with Ionic Liquid Pendants. *J. Am. Chem. Soc.* **2009**, *131*, 17722–17723. [CrossRef] [PubMed]
37. Li, C.H.; He, J.H.; Liu, J.H.; Qian, L.A.; Yu, Z.Q.; Zhang, Q.L.; He, C.X. Liquid Crystalline Phases of 1,2-Dimethyl-3-Hexadecylimidazolium Bromide and Binary Mixtures With Water. *J. Colloid Interface Sci.* **2010**, *349*, 224–229. [CrossRef]

38. Goossens, K.; Wellens, S.; Van Hecke, K.; Van Meervelt, L.; Cardinaels, T.; Binnemans, K. T-Shaped Ionic Liquid Crystals Based on the Imidazolium Motif: Exploring Substitution of the C-2 Imidazolium Carbon Atom. *Chem. Eur. J.* **2011**, *17*, 4291–4306. [CrossRef]

39. Fernandez, A.A.; de Haan, L.T.; Kouwer, P.H.J. Towards Room-Temperature Ionic Liquid Crystals. *J. Mater. Chem. A* **2013**, *1*, 354–357. [CrossRef]

40. Nestor, S.T.; Heinrich, B.; Sykora, R.A.; Zhang, X.; McManus, G.J.; Douce, L.; Mirjafari, A. Methimazolium-Based Ionic Liquid Crystals: Emergence of Mesomorphic Properties via a Sulfur Motif. *Tetrahedron* **2017**, *73*, 5456–5460. [CrossRef]

41. Hindman, M.S.; Stanton, A.D.; Christopher Irvin, A.; Wallace, D.A.; Moon, J.D.; Reclusado, K.R.; Liu, H.; Belmore, K.A.; Liang, Q.; Shannon, M.S.; et al. Synthesis of 1,2-Dialkyl-, 1,4(5)-Dialkyl-, and 1,2,4(5)-Trialkylimidazoles via a One-Pot Method. *Ind. Eng. Chem. Res.* **2013**, *52*, 11880–11887. [CrossRef]

42. Yue, S.; Roveda, J.D.; Mittenthal, M.S.; Shannon, M.S.; Bara, J.E. Experimental Densities and Calculated Fractional Free Volumes of Ionic Liquids With Tri- and Tetra-Substituted Imidazolium Cations. *J. Chem. Eng. Data* **2018**, *63*, 2522–2532. [CrossRef]

43. Fumino, K.; Wulf, A.; Ludwig, R. Strong, Localized, and Directional Hydrogen Bonds Fluidize Ionic Liquids. *Angew. Chem. Int. Ed.* **2008**, *47*, 8731–8734. [CrossRef] [PubMed]

44. Fumino, K.; Wulf, A.; Ludwig, R. The Potential Role of Hydrogen Bonding in Aprotic and Protic Ionic Liquids. *Phys. Chem. Chem. Phys.* **2009**, *11*, 8790–8794. [CrossRef] [PubMed]

45. Roth, C.; Peppel, T.; Fumino, K.; Köckerling, M.; Ludwig, R. The Importance of Hydrogen Bonds for the Structure of Ionic Liquids: Single-Crystal X-ray Diffraction and Transmission and Attenuated Total Reflection Spectroscopy in the Terahertz Region. *Angew. Chem. Int. Ed.* **2010**, *49*, 10221–10224. [CrossRef] [PubMed]

46. Wulf, A.; Fumino, K.; Ludwig, R. Spectroscopic Evidence for an Enhanced Anion-Cation Interaction from Hydrogen Bonding in Pure Imidazolium Ionic Liquids. *Angew. Chem. Int. Ed.* **2010**, *49*, 449–453. [CrossRef] [PubMed]

47. Peppel, T.; Roth, C.; Fumino, K.; Paschek, D.; Köckerling, M.; Ludwig, R. The Influence of Hydrogen-Bond Defects on the Properties of Ionic Liquids. *Angew. Chem. Int. Ed.* **2011**, *50*, 6661–6665. [CrossRef] [PubMed]

48. Fumino, K.; Peppel, T.; Geppert-Rybczynska, M.; Zaitsau, D.H.; Lehmann, J.K.; Verevkin, S.P.; Köckerling, M.; Ludwig, R. The Influence of Hydrogen Bonding on the Physical Properties of Ionic Liquids. *Phys. Chem. Chem. Phys.* **2011**, *13*, 14064–14075. [CrossRef]

49. Hugar, K.M.; Kostalik, H.A.; Coates, G.W. Imidazolium Cations with Exceptional Alkaline Stability: A Systematic Study of Structure-Stability Relationships. *J. Am. Chem. Soc.* **2015**, *137*, 8730–8737. [CrossRef]

50. Wang, D.; Richter, C.; Rühling, A.; Drücker, P.; Siegmund, D.; Metzler-Nolte, N.; Glorius, F.; Galla, H.-J. A Remarkably Simple Class of Imidazolium-Based Lipids and Their Biological Properties. *Chem. Eur. J.* **2015**, *21*, 15123–15126. [CrossRef]

51. Rühling, A.; Wang, D.; Ernst, J.B.; Wulff, S.; Honeker, R.; Richter, C.; Ferry, A.; Galla, H.-J.; Glorius, F. Influence of the Headgroup of Azolium-Based Lipids on Their Biophysical Properties and Cytotoxicity. *Chem. Eur. J.* **2017**, *23*, 5920–5924. [CrossRef]

52. Drücker, P.; Rühling, A.; Grill, D.; Wang, D.; Draeger, A.; Gerke, V.; Glorius, F.; Galla, H.-J. Imidazolium Salts Mimicking the Structure of Natural Lipids Exploit Remarkable Properties Forming Lamellar Phases and Giant Vesicles. *Langmuir* **2017**, *33*, 1333–1342. [CrossRef] [PubMed]

53. Rakers, L.; Glorius, F. Flexible Design of Ionic Liquids for Membrane Interactions. *Biophys. Rev.* **2018**, *10*, 747–750. [CrossRef] [PubMed]

54. Rondla, R.; Lee, C.K.; Lu, J.T.; Lin, I.J.B. Symmetrical 1,3-Dialkylimidazolium Based Ionic Liquid Crystals. *J. Chin. Chem. Soc.* **2013**, *60*, 745–754.

55. Wang, X.; Heinemann, F.W.; Yang, M.; Melcher, B.U.; Fekete, M.; Mudring, A.-V.; Wasserscheid, P.; Meyer, K. A New Class of Double Alkyl-Substituted, Liquid Crystalline Imidazolium Ionic Liquids—A Unique Combination of Structural Features, Viscosity Effects, and Thermal Properties. *Chem. Commun.* **2009**, 7405–7407. [CrossRef] [PubMed]

56. Wang, X.; Vogel, C.S.; Heinemann, F.W.; Wasserscheid, P.; Meyer, K. Solid-State Structures of Double-Long-Chain Imidazolium Ionic Liquids: Influence of Anion Shape on Cation Geometry and Crystal Packing. *Cryst. Growth Des.* **2011**, *11*, 1974–1988. [CrossRef]

57. Wang, X.; Sternberg, M.; Kohler, F.T.U.; Melcher, B.U.; Wasserscheid, P.; Meyer, K. Long-Alkyl-Chain-Derivatized Imidazolium Salts and Ionic Liquid Crystals with Tailor-Made Properties. *RSC Adv.* **2014**, *4*, 12476–12481. [CrossRef]

58. Yang, M.; Mallick, B.; Mudring, A.-V. A Systematic Study on the Mesomorphic Behavior of Asymmetrical 1-Alkyl-3-Dodecylimidazolium Bromides. *Cryst. Growth Des.* **2014**, *14*, 1561–1571. [CrossRef]

59. Lee, C.K.; Peng, H.H.; Lin, I.J.B. Liquid Crystals of *N,N'*-Dialkylimidazolium Salts Comprising Palladium(II) and Copper(II) Ions. *Chem. Mater.* **2004**, *16*, 530–536. [CrossRef]

60. Ngo, H.L.; LeCompte, K.; Hargens, L.; McEwen, A.B. Thermal Properties of Imidazolium Ionic Liquids. *Thermochim. Acta* **2000**, *357*, 97–102. [CrossRef]

61. Park, G.; Goossens, K.; Shin, T.J.; Bielawski, C.W. Dicyanamide Salts That Adopt Smectic, Columnar, or Bicontinuous Cubic Liquid-Crystalline Mesophases. *Chem. Eur. J.* **2018**, *24*, 6399–6411. [CrossRef]

62. Marcos, M.; Giménez, R.; Serrano, J.L.; Donnio, B.; Heinrich, B.; Guillon, D. Dendromesogens: Liquid Crystal Organizations of Poly(Amidoamine) Dendrimers Versus Starburst Structures. *Chem. Eur. J.* **2001**, *7*, 1006–1013. [CrossRef]

63. Fouchet, J.; Douce, L.; Heinrich, B.; Welter, R.; Louati, A. A Convenient Method for Preparing Rigid-Core Ionic Liquid Crystals. *Beilstein J. Org. Chem.* **2009**, *5*, No. 51. [CrossRef] [PubMed]

64. Dobbs, W.; Douce, L.; Heinrich, B. 1-(4-Alkyloxybenzyl)-3-methyl-1*H*-imidazol-3-ium Organic Backbone: A Versatile Smectogenic Moiety. *Beilstein J. Org. Chem.* **2009**, *5*. No. 62. [CrossRef] [PubMed]

65. Jenkins, H.D.B.; Roobottom, H.K.; Passmore, J.; Glasser, L. Relationships among Ionic Lattice Energies, Molecular (Formula Unit) Volumes, and Thermochemical Radii. *Inorg. Chem.* **1999**, *38*, 3609–3620. [CrossRef] [PubMed]

66. Dzyuba, S.V.; Bartsch, R.A. New Room-Temperature Ionic Liquids with C_2-Symmetrical Imidazolium Cations. *Chem. Commun.* **2001**, *16*, 1466–1467. [CrossRef]

67. Goossens, K.; Nockemann, P.; Driesen, K.; Goderis, B.; Görller-Walrand, C.; Van Hecke, K.; Van Meervelt, L.; Pouzet, E.; Binnemans, K.; Cardinaels, T. Imidazolium Ionic Liquid Crystals With Pendant Mesogenic Groups. *Chem. Mater.* **2008**, *20*, 157–168. [CrossRef]

68. Tschierske, C. Liquid Crystal Engineering—New Complex Mesophase Structures and Their Relations to Polymer Morphologies, Nanoscale Patterning and Crystal Engineering. *Chem. Soc. Rev.* **2007**, *36*, 1930–1970. [CrossRef] [PubMed]

69. Ichikawa, T.; Yoshio, M.; Hamasaki, A.; Taguchi, S.; Liu, F.; Zeng, X.; Ungar, G.; Ohno, H.; Kato, T. Induction of Thermotropic Bicontinuous Cubic Phases in Liquid-Crystalline Ammonium and Phosphonium Salts. *J. Am. Chem. Soc.* **2012**, *134*, 2634–2643. [CrossRef] [PubMed]

70. Ungar, G.; Liu, F.; Zeng, X. Cubic and Other 3D Thermotropic Liquid Crystal Phases and Quasicrystals. In *Handbook of Liquid Crystals. Volume 5: Non-Conventional Liquid Crystals*, 2nd ed.; Goodby, J.W., Collings, P.J., Kato, T., Tschierske, C., Gleeson, H., Raynes, P., Eds.; Wiley-VCH: Weinheim, Germany, 2014; pp. 363–436.

crystals

MDPI

Article

Phase Behaviors of Ionic Liquids Heating from Different Crystal Polymorphs toward the Same Smectic-A Ionic Liquid Crystal by Molecular Dynamics Simulation

Wudi Cao [1,2] and Yanting Wang [1,2,*]

[1]   CAS Key Laboratory of Theoretical Physics, Institute of Theoretical Physics, Chinese Academy of Sciences, 55 East Zhongguancun Road, P.O. Box 2735, Beijing 100190, China; caowudi@itp.ac.cn
[2]   School of Physical Sciences, University of Chinese Academy of Sciences, 19A Yuquan Road, Beijing 100049, China
*   Correspondence: wangyt@itp.ac.cn

Received: 11 December 2018; Accepted: 29 December 2018; Published: 3 January 2019

Abstract: Five distinct crystal structures, based on experimental data or constructed manually, of ionic liquid [C_{14}Mim][NO_3] were heated in *NPT* molecular dynamics simulations under the same pressure such that they melted into the liquid crystal (LC) phase and then into the liquid phase. It was found that the more entropy-favored structure had a higher solid-LC transition temperature: Before the transition into the LC, all systems had to go through a metastable state with the side chains almost perpendicular to the polar layers. All those crystals finally melted into the same smectic-A LC structure irrelevant of the initial crystal structure.

Keywords: ionic liquid; phase behavior; crystal polymorphs; ionic liquid crystal

---

1. Introduction

Ionic liquid crystals (ILCs) are the intersections of ionic liquids (ILs) and liquid crystals (LCs), with the features of ILs being comprised of pure ions and of LCs having order(s) in one or two dimensions, whose application prospects [1,2] have been appreciated in the most recent twenty years.

Long-chain imidazolium-based ILs [C_nMim]X, with a proper combination of alkyl-chain length *n* and anion type X, are well known to be able to form LC structures and have been intensively studied. The combinations have been experimentally determined to be $n \geq 10$ for (FH)$_2$F, $n \geq 12$ for NO_3 and BF_4, $n \geq 14$ for PF_6, etc. [3]. Their thermotropic phase behavior, as an essential physical property, has been investigated by both experiments [3–12] and simulations [13–17]. In experiments, IL samples usually undergo several heating-cooling cycles, heated from a crystal phase to reach LC and/or liquid phases and then cooled down to recrystallize, during which crystal polymorphism is often reported, which is potentially of great importance in materials science and industry because physiochemical properties of polymorphs, such as thermal stability [6,9,18], ion conductivity [9,12], etc., can significantly affect productions and applications of ILs and ILCs. Due to the complexity of IL and ILC systems, however, the microscopic details of the phase behaviors related to IL polymorphs can only be speculative solely based on experimental data, in which molecular dynamics (MD) simulations can play a vital role [14,18,19].

From a simulation viewpoint, there are two convenient ways to obtain ILC structures. One is from heating a crystal structure [13–15,20–22], and another is through simulated annealing processes [17,23]. For the first way, the initial crystal structures are either handmade or from experimental data and may differ from one work to another, exhibiting the existence of crystal polymorphs. If the unit cells of these polymorphs are very different, the transitions between them can hardly be observed during normal MD simulations without specifically designed techniques [24–27]. In other words, when simply heating simulation systems with initial crystal structures significantly different from each other, they go through different pathways in the phase space toward melting. The relation between the microscopic details and pathways has rarely been studied before, and the temperature at which the pathways join up is unclear. In addition, the structure of ILCs might in principle depend on thermal history, so it is necessary to clarify whether ILCs obtained from different crystal polymorphs are the same, i.e., whether the heating pathways merge above the solid-ILC phase transition temperatures.

In our previous simulations [14], we began with manually constructed crystal structures of ILs [C_nMim][NO_3] to study the influence of alkyl-chain length on phase behavior during heating. Initial crystal structures for different systems with chain lengths of n = 4, 6, 8, 10, 12 were prepared as similarly as possible. During heating, a solid-solid phase transition took place for all the systems, and a metastable state occurred at the melting point for all except n = 6. The ILC structure was observed for n = 10 and 12, but not for smaller values of n.

On the basis of the previous discoveries, in this work, we prepared five crystal polymorphs of [C_{14}Mim][NO_3], focusing on the evolution of structures with temperature. Three of the polymorphs were based on experimentally determined structures of [C_{14}Mim]X [9,28], where anions X were replaced by nitrates, and the other two were manually constructed and recursively refined. The anions of the polymorphs were the same to exclude the impact of different anion types [7,29,30], and the choice of nitrate enabled a direct comparison to previous results [13,14,17,20–23,31,32]. With these crystals as initial structures, we performed a series of *NPT* MD simulations with the temperature being gradually increased. It has been observed that all the initial structures go through solid-LC and LC-liquid phase transitions sequentially, ending with a nanoscale segregation liquid (NSL) structure [31–33]. The potential energy, density, layer spacing, and some order parameters were computed to locate transition points, characterize structural features, and recognize phases. Different crystal polymorphs take different pathways to reach the ILC state at distinct solid-LC phase transition temperatures. According to the simulation data, we propose that the [C_{14}Mim][NO_3] IL is a special enantiotropic system, the schematic of which is compared to that of a monotropic system in Figure 1, where CrI and CrII refer to any two of the five crystal polymorphs. As shown in Figure 1a, in a common enantiotropic system, the low-temperature crystal CrI transits to a high-temperature crystal CrII, and the system melts at the melting point of CrII, but this solid-solid phase transition did not take place for the [C_{14}Mim][NO_3] IL during our MD simulations, leading to an interesting phenomenon that CrI and CrII had two distinct melting points, so the system behaved somewhat like a monotropic system. However, in a monotropic system (Figure 1b), an entropy-favored, disordered structure has a lower melting point, but here the entropy-favored structure had a higher one, so we conclude that [C_{14}Mim][NO_3] is a special enantiotropic system. During solid-LC transitions, what always presents is a metastable state featuring cationic alkyl-chains being perpendicular to polar layers, as already observed in our previous simulations [14]. The metastable state eventually evolved into the same smectic-A (SmA) LC structure, irrelevant of the initial crystalline structure and heating pathway. The simulation temperatures kept increasing until the LC-liquid phase transitions happened. The potential energy changes at the clearing point were much smaller than those at the melting points, which agreed with the experimental results of ILCs composed of the same imidazolium cation and different anions [4–7,10,12].

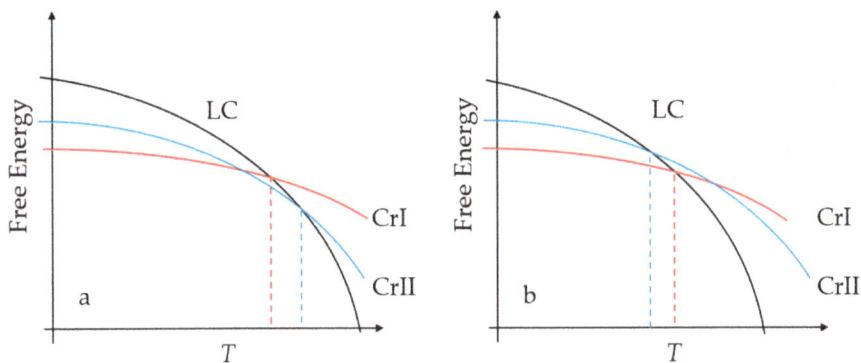

Figure 1. Schematics of (**a**) enantiotropic and (**b**) monotropic systems. CrI and CrII are two crystal polymorphs. We propose that the ionic liquid (IL) polymorphism in this work belongs to the former, but the transition from CrI to CrII did not occur during our molecular dynamics (MD) simulations.

2. Methods

2.1. Simulation Setup

As we did in our previous works [14], we adopted an all-atom model based on an AMBER force field [34]. To obtain partial charges, we first performed an ab initio optimization at the mp2/6-31g* level with a Gaussian package [35], and then produced charges by fitting the electrostatic potential in space according to the Merz–Singh–Kollman scheme [36,37] at the same level. The partial charges were finally determined by the RESP utility [38] provided in the AMBER package. The MD simulations were run with the GROMACS 5.1.4 package [39]. The systems were coupled with an anisotropic or semi-isotropic Parrinello–Rahman barostat [40] to keep the shape of the simulation box flexible and properly commensurate with the crystal and LC structures. Each component of the reference pressure was set to 1 bar, and the compressibility was 4.5×10^{-5} bar^{-1}. The temperature was controlled by a Nosé–Hoover thermostat [41], and a periodic boundary condition (PBC) was applied to all three dimensions. Electrostatic interactions were computed with the Particle-Mesh Ewald method [42] with a real-space cutoff of 12 Å, the same as the cutoff for van der Waals (VDW). VMD software [43] was used when visualizing simulation trajectories and when constructing and modifying crystal structures.

In the production simulations, temperatures were increased stepwise with a step of 25 K (50 K for the larger-size double bilayer (DB) system described below), and 5 K near the transition points. For the crystal phase, whose fluctuations were small, the systems were sampled within 10 ns. When the temperature approached the melting point and the systems became LCs or liquid structures whose fluctuations became large, the systems were sampled for more than 10 ns and up to 40 ns when necessary.

2.2. Preparation of Initial Configurations

The first handmade crystal structure, denoted by HM1 hereafter, was obtained by elongating the alkyl chains of the manually constructed [C$_{12}$Mim][NO$_3$] structure we presented in a previous work [14]. The second handmade one (HM2) was newly constructed by greatly enlarging the tilt angle of the alkyl chains in HM1, so the layer spacing became much narrower. The other three structures were obtained by modifying the experimentally determined structures by replacing the anions with nitrates and removing the water molecules wherever they presented. DB represents the double bilayer structure modified from Reference [28], whereas EM1 and EM2 refer to the bilayer structures modified from References [28,29], respectively. The simulation boxes were made up by copying unit cells along the lattice vectors. The lattice vector a was always along the x axis, and b was in the x–y plane, so the polar bilayers were parallel to the x–y plane. The unit cells of HM1 and HM2 with 4 ion pairs were copied $6 \times 6 \times 3$ times along lattice vectors a, b, and c, so the simulation boxes totally contained 432 ion pairs. A unit cell of a DB with 6 ion pairs was copied $4 \times 4 \times 4$ times and $6 \times 6 \times 3$ times to form two system sizes of 384 and 684 ion pairs, respectively. EM1, having 2 ion pairs per unit cell, was copied $6 \times 4 \times 8$ times, and EM2, with 4 ion pairs, was copied $4 \times 3 \times 8$ times, both resulting in a simulation box with 384 ion pairs.

With the anisotropic barostat applied, the crystal lattices were equilibrated at 5 K for 200 ps. Series of 2-ns *NPT* simulations then followed to increase the temperature stepwise to 300 K for further equilibration. The systems were annealed to 200 K, at which the production simulations began. The obtained crystal data of the polymorphs at 200 K are listed in Table 1, and the unit cells are drawn in Figure 2. Although the anions were replaced by nitrates, the DB, EM1, and EM2 retained their essential structural features with the space groups unchanged. The conformation of cation was also preserved: The cations were straight in HM1, HM2, and EM1; straight in one layer and bend (head group perpendicular to side chain) in neighboring layers in the DB; and "crank-handle"-like in EM2.

Table 1. Lattice constants of the five polymorphs at T = 200 K.

Crystal Data	HM1	HM2	DB	EM1	EM2
a (Å)	9.27	13.22	7.79	6.50	7.75
b (Å)	8.00	8.19	8.16	6.65	9.80
c (Å)	27.21	20.15	52.01	22.74	25.81
α (°)	84.3	78.1	84.5	91.2	90.0
β (°)	86.1	103.5	89.3	96.1	86.7
γ (°)	74.3	73.9	63.2	100.0	90.1
Z	4	4	6	2	4
Crystal system	Triclinic	Triclinic	Triclinic	Triclinic	Monoclinic
Space group	$P\bar{1}$	$P\bar{1}$	$P\bar{1}$	$P\bar{1}$	$P2_1/a$
ρ (g/cm^3)	1.174	1.164	1.157	1.177	1.158

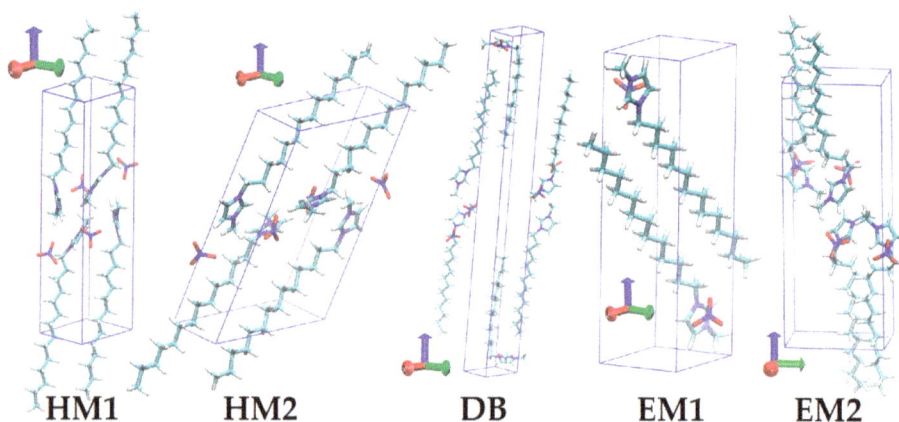

Figure 2. Unit cells of crystal structures at T = 200 K. The lattice vector a is along the x axis (red).

2.3. Order Parameters

Different phases have different spatial orders, which can be quantified by carefully defined order parameters. Distinct phases can be identified by a combination of several order parameters. To quantify the translational order of crystal structures, the translational order parameter (TOP) of a configuration is defined as

$$\tau = \frac{1}{n_t} \sum_{\alpha=0}^{n_t} \left| \frac{1}{N_{UC}} \sum_{j=0}^{N_{UC}} \exp(i\vec{r_j^\alpha} \cdot \vec{G}) \right|, \tag{1}$$

where vector r denotes the position of the αth site in the jth unit cell, vector G is the sum of the three basis vectors of reciprocal lattice, $| \ldots |$ means taking the modulus, and n_t and N_{UC} are the number of a certain type of sites in the unit cell and the number of unit cells in the simulation box, respectively.

When the systems are in the SmA LC phase, which has no crystal lattices but a one-dimensional translational periodicity, the TOP is modified accordingly as [44]

$$\tau_{LC} = \frac{1}{N_t} \sum_{\alpha=0}^{N_t} |\exp(iz^{\alpha}/d)|, \tag{2}$$

where z is the z-component of the position, d is the layer spacing, and N_t is the total number of cationic headgroups and anions in the simulation box.

The basis vectors of the reciprocal lattice are determined from the real-space lattice. One set of the basis vectors of the real-space lattice is obtained by dividing the box vectors by the numbers of copies of unit cells along them. One of the basis vectors in the set is fixed in GROMACS to be along the x axis, and the other two vectors are further selected among the allowed primitive vectors (linear combinations of the other two basis vectors in the set) so that τ takes the maximum. Similarly, for LCs, the layer spacing d is chosen by maximizing τ_{LC}.

With a directional vector defined for a cation connecting the center of masses (COMs) of its head and tail groups (the CH_3 group at the end of alkyl chain), the orientational order parameter (OOP) of a configuration is defined to quantify the orientational order of the system, which is

$$S = \langle P_2(\cos\theta) \rangle, \tag{3}$$

where P_2 denotes the third term of the Legendre polynomials, θ is the angle between the directional vector of a cation and the average directional vector of the simulation box, and $\langle \ldots \rangle$ means to take the average among the cations.

To quantify the conformational order of alkyl chains, all the dihedral angles defined by three consecutive carbon–carbon bonds on the same chain are computed, based upon which the ratio of the *trans*-conformation, whose angle is between $150°$ and $180°$, is calculated and used as the conformational order parameter, denoted by R_t.

These order parameters, together with the energy and density of the systems, are taken into account when judging whether the system is in equilibrium or at least in local equilibrium. The averages and variances of these quantities over nanosecond time scales are computed when their values with respect to time have no evident drift.

3. Results

When heating the systems, all the initial crystal polymorphs underwent phase transitions from crystal to LC and then from LC to liquid. Solid-solid phase transitions could happen below the solid-LC transition temperature, and a metastable state always occurred during the solid-LC transition. Note that the solid-solid phase transitions were not from one of the five initial structures to another, but rather the crystal structures initiated from the five simulated systems before melting were always different from each other. We present the snapshots of these phases during the simulations of HM1 in Figure 3. The snapshots of the metastable state, LCs, and NSL for the other four structures are basically the same. In the crystal phase (Figure 3a), all the polymorphs shared the same feature that cationic headgroups and anions arranged in an ordered pattern, forming polar bilayers with the nonpolar region composed of all-*trans* interdigital alkyl chains in between (except that EM2 had a gauche configuration near the headgroup). This structural feature held for the whole crystal range despite the fact that the lattice constants differed from polymorph to polymorph and changed during the solid-solid phase transitions. The DB structure was exceptional in that the number densities of cationic headgroups and anions in neighboring polar bilayers were different (see the unit cell in Figure 2): The layer with straight cations had a number density twice as much as the layer with headgroups bent to the alkyl chains, resulting in a doubled periodicity. In the metastable state (Figure 3b) between crystal and LC, which had an intermediate order, the alkyl chains turned to be more perpendicular to the polar bilayers and no longer in the straight all-*trans* style, but they were still interdigital with each other. The LC structure (Figure 3c) was even more disordered than the metastable structure, although the structural layering still existed, whose alkyl chains were no longer interdigitally aligned due to large thermal energy, and thus the layer spacing increased accordingly. As the temperature further went up, the polar layers dispersed, and the system eventually changed into the NSL phase (Figure 3d).

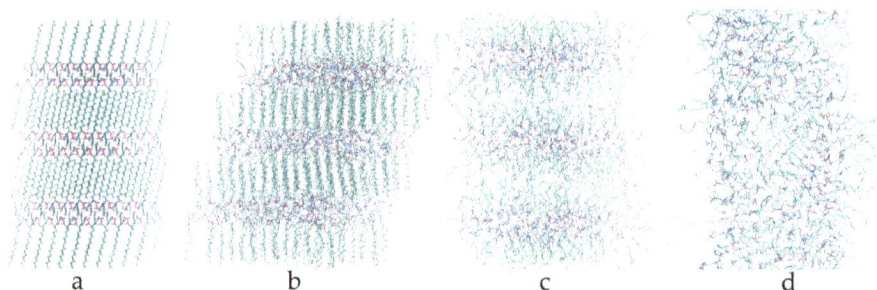

| a | b | c | d |

Figure 3. Snapshots of $[C_{14}Mim][NO_3]$ beginning with the HM1 structure at various temperatures. (**a**) HM1 crystal at 200 K; (**b**) Metastable state during the solid-liquid crystal (LC) transition at 455 K; (**c**) SmA LC at 550 K; (**d**) Nanoscale segregation liquid (NSL) at 650 K.

The transition temperatures and corresponding potential energy changes are listed in Table 2. The details of the structures appearing in the table could be characterized by the density, layer spacing, and order parameters, as is shown below. The phase transitions in Table 2 corresponded to the kinks on the caloric curves shown in Figure 4. The *PV* term was in fact negligible (~$10-2$ kJ/mol) compared to the large potential energy since the interactions in our model were strong and the applied pressure (1 bar) was relatively small, so the potential energy changes basically equaled the enthalpy changes. Note that in EM2, there were two solid-solid phase transitions, whose potential energy changes were both negative because the initial structure obtained by modifying the reported structure of [C14Mim][PF6] and re-equilibrating below 300 K was not the most stable structure for [C14Mim][NO3], and when the temperature was adequately high, the structure could transit into the more stable structure with a lower potential energy. Moreover, the LC-liquid transitions ca. 655 K were unidentifiable only by looking at the caloric curves in Figure 4, since the potential energy differences were too small, which agreed well with the experimental results of similar ILCs [4–7,10,12].

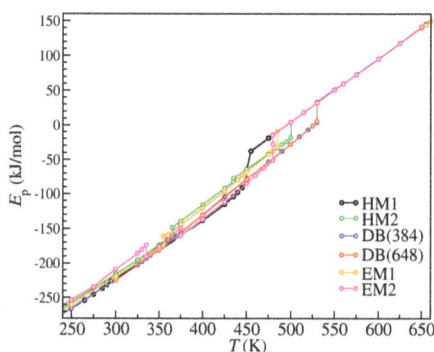

Figure 4. Caloric curve of one mole $[C_{14}Mim][NO_3]$ in each system during heating.

Table 2. Phase transition temperatures and potential energy changes.

Initial Crystal Polymorph	Phase Transition	T (K)	ΔE_p (kJ/mol)
HM1	Solid-metastable	450	18.92
	Metastable-LC	455	22.45
	LC-liquid	650	1.98
HM2	Solid-solid	365	9.52
	Solid-solid	435	4.71
	Solid-LC	500	23.04
	LC-liquid	655	1.58
DB	Solid-LC	530	29.52
	LC-liquid	655	1.58
EM1	Solid-solid	355	12.31
	Solid-LC	485	22.85
	LC-liquid	660	0.34
EM2	Solid-solid	335	−16.29
	Solid-solid	375	−11.27
	Solid-metastable	480	23.21
	Metastable-LC	480	13.55
	LC-liquid	650	2.07

The structural differences between polymorphs and the structural changes during phase transitions could be illustrated by density and layer spacing, shown in Figure 5. The layer spacing d in Equation (2) of a crystal is the height of a unit cell perpendicular to the polar bilayers, and that of an LC is the one-dimensional periodicity along the ordered direction. The density reflected the compactness of packing, and no experimental density data for [C$_{14}$Mim][NO$_3$] were available for comparison. Figure 5 shows that the crystal polymorphs had similar densities, but different layer spaces. Sharp changes in density and layer spacing took place when the solid-solid and solid-LC phase transitions happened. In contrast, for the LC-liquid transition, the density changes were very small, and the fluctuations of the estimated LC layer spacing increased with temperature and became divergent at the clearing point.

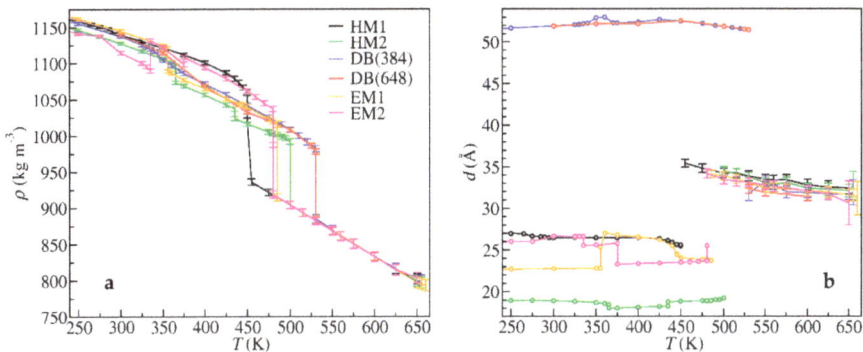

Figure 5. (**a**) System density and (**b**) layer spacing versus temperature. The error bars represent variances, and for the points without error bars, the error bars are smaller than the symbols.

The crystal phase was quantitatively characterized by computing the translational order parameters (TOPs). The crystalline TOP computed for the COMs of cations according to Equation (1) are shown in Figure 6a. For the HM1 and HM2 structures with 432 ion pairs, the TOPs kept above ca. 0.95 until they melted into LCs, justifying that they were strict in crystalline arrangement throughout the crystal phase. For the DB, EM1, and EM2 with 384 ion pairs, the TOPs were also beyond 0.7 before melting, although the values were relatively smaller, and the fluctuations were large. However, a series of additional simulations of the DB system with 648 ion pairs showed larger TOPs and smaller fluctuations, as can be seen in Figure 6b, indicating a noticeable finite-size effect. Nevertheless, the TOP values for the system with 384 ion pairs were already large enough to demonstrate that the COMs of cations were basically at the lattice positions before the melting points.

Similarly, the SmA LC was quantified by one-dimensional LC TOP, defined in Equation (2), calculated for the COMs of cationic headgroups and the COMs of anions. The LC TOP is also shown in Figure 6a. Compared to the large values of crystalline TOPs, the LC TOP was only about 0.5, indicating that the LC phase of the IL was rather disordered and merely had a rough layered structure.

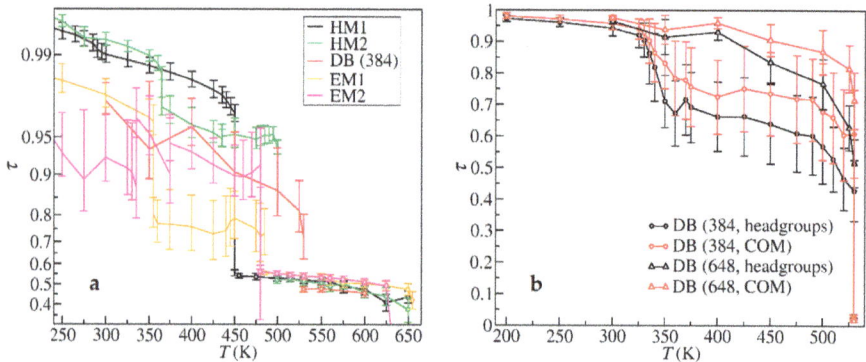

Figure 6. (**a**) Translational order parameters (TOPs) of the crystal and SmA liquid crystal structures; (**b**) Crystalline TOPs of the DB structure calculated for the center of masses (COMs) of headgroups, anions, and cations with 384 and 648 ion pairs, respectively.

The orientational order parameters (OOPs) and the number ratios of the *trans* conformation of the alkyl chains, defined in Equations (2) and (3), respectively, are shown in Figure 7. As indicated by the OOP values shown in Figure 7a, which were close to 1 at low temperatures, the cations in the unit cell were almost along the same direction. The OOPs decreased sharply to about 0.5 when the crystals turned to LCs, suggesting that, unlike the common rod-like LCs, the soft alkyl chains of ILCs were not well aligned. When the temperature increased to around 655 K, the OOPs decreased suddenly to 0.1~0.2, corresponding to the occurrence of the LC-liquid transitions. The OOP values did not go to zero probably due to the finite-size effect.

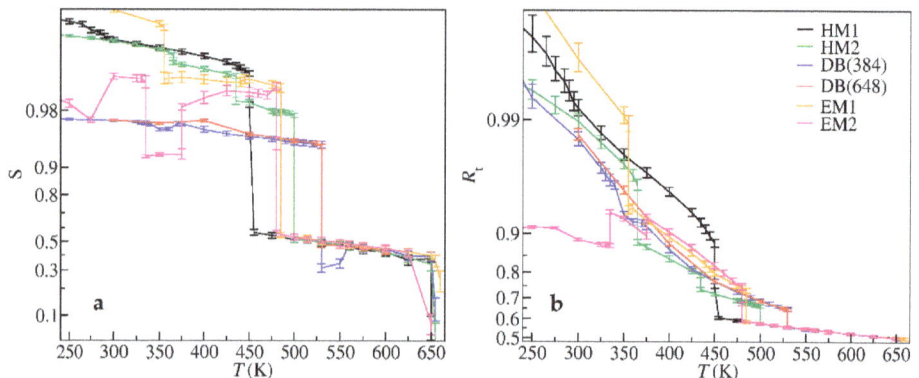

Figure 7. (**a**) Orientational order parameters (OOPs) and (**b**) number ratios of *trans* conformations versus temperature during heating.

As for the number ratio of *trans* conformations, Figure 7b shows that the alkyl chains in the crystal phase, especially at low temperatures, were basically in the all-*trans* conformation. The ratio of EM2 crystal was an exception, and was about 0.909 (10/11), because of the gauche conformation composed of the first four carbons near the headgroup, i.e., the "crank-handle"-like conformation (see the unit cell shown in Figure 2). As the temperature increased, solid-solid phase transitions happened, and the alkyl chain conformations became less ordered, indicated by the decrease of the *trans*-conformation ratio. At the melting point, the decrease was more evident, because the breakdown of the interdigital structures allowed the alkyl chains to take more conformations without strong constraints between each other. On the contrary, the *trans*-conformation ratio remained unchanged at the clearing point, demonstrating that the conformations of alkyl chains in LC and in liquid were the same.

4. Discussion

It should be noted that, as a common problem for MD simulations, the transition temperatures were different from those obtained from experiments [3] because of hysteresis. It is a knotty problem to acquire transition temperatures in MD simulations comparable to experimental results for complex systems such as ILs, even with the aid of some free-energy-involved methods [45–47]. Nevertheless, the transition temperatures for all systems were systematically shifted to one side in the MD simulations, which did not hinder the qualitative comparison with each other.

There are two types of crystal polymorphisms, which are schematically shown in Figure 1. For the enantiotropic system, solid-solid phase transitions from one to another take place below the melting points, but not for the monotropic system. We suggest that the IL system in this work is an enantiotropic system. Note that the free energy at zero temperature equals the potential energy, and the slopes of the curves in Figure 1 equal the opposite of entropies. If CrI does not turn to CrII at the intersection of their free energy curves in Figure 1a in a finite-time simulation, it turns to LC at a lower temperature with a smaller entropy near the transition points, and CrII turns to LC at a higher temperature with a larger entropy, which coincides with the simulation data of the IL system here: Generally, the IL system with a lower density (Figure 5a) and smaller order parameters (Figure 7), indicating a more disordered structure with a larger entropy, tended to have a higher solid-LC transition point.

In our MD simulations, CrI and CrII could refer to any two of the five polymorphs, and we did not see the solid-solid phase transition from CrI to CrII, which should thermodynamically happen according to the diagram in Figure 1a. The reason was that the energy barrier between the two crystal structures was so high that this was almost impossible to occur during very limited MD simulation time and even experimental time [7]. Moreover, the type of anion may also have affected the occurrence of the solid-solid phase transition, as it has been reported to happen in ILs with BF_4 [12], but not with other anions. After all, the inhibition of transitions between different crystal structures leads to the counterintuitive phenomenon that more entropy-favored systems have higher solid-LC transition temperatures.

The processes of the solid-LC transitions were studied in detail. First, for the metastable states occurring during this transition, we could see by eye that the alkyl chains became upright. We therefore calculated the orientation of the cations to quantify this phenomenon and show the results in Figure 8.

As shown in Figure 8a, during the solid-LC transition, the system first turned to a metastable structure whose lifetime could be very short, and the alkyl chains became perpendicular to the polar layers, before it melted into the SmA LC, except the case of the DB, when the alkyl chains were already perpendicular in the crystal structure. Take the example of EM2, whose metastable state had a relatively long lifetime. From Figure 8b, in the crystal phase, both the polar and azimuth angles of alkyl chains were restricted in a narrow range. In the metastable state, the polar angles were still restricted to ca. 10°, but the azimuth angles were evenly distributed in all directions, demonstrating that the alkyl chains were on average perpendicular to the polar layers. In the SmA LC, the polar angles increased to ca. 30° with a broader distribution because the alkyl chains were no longer interdigital and bent. From a thermodynamic viewpoint, the distributions of alkyl-chain directions in metastable states were more symmetric than those in crystals, and the *trans*-conformation ratio of alkyl chains in metastable states was also smaller, leading to a larger entropy than crystal, but still smaller than LC. On the other hand, the alkyl chains in metastable state were still in the interdigital style, so the VDW potential between the alkyl chains, and thus the potential energy of the system, did not increase a lot. Together, the structure of metastable states favored entropy more without a large structural change costing the potential energy too much, so it was a reasonable intermediate structure along the solid-LC transition pathway.

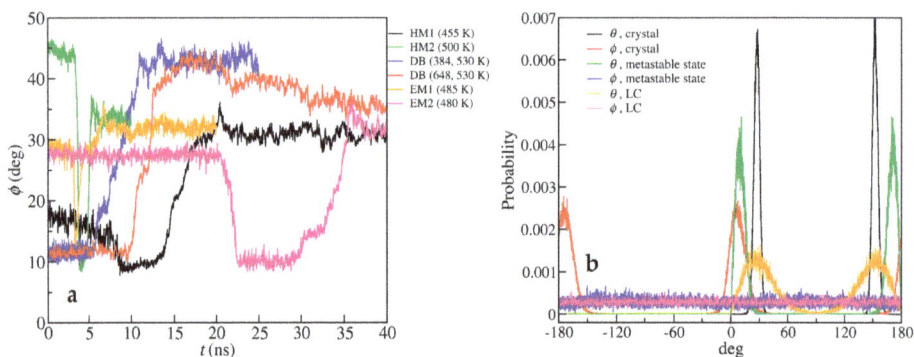

Figure 8. (a) The average angle between cations and the z axis ranging from 0° to 90° evolving with time at the solid-LC transition temperature; (b) The angle distributions of cation orientation vectors in different stages of solid-LC transitions of EM2 at 480 K, where θ is the polar angle ranging from 0° to 180°, and ϕ is the azimuth angle ranging from −180° to 180°.

All the above quantities clearly show that, after the solid-LC transitions, all these crystal polymorphs transformed into the same SmA ILC. The curves for those quantities, which were separated from each other in the crystal ranges, overlapped very well after the solid-LC transition. This was true even for the DB structure with alternate anion densities in polar layers, as shown in Figure 9. When the crystal phase transitioned to the LC phase, the layer spacing increased, and the headgroups and anions moved across the nonpolar regions to rearrange, leading to nonzero densities of ions in nonpolar regions and then uniform polar layers with the same density. In the larger-size DB system (Figure 9b), this rearrangement completed shortly at the melting point of 530 K, with the anion densities in nonpolar regions coming back to zero, but it was more difficult for the smaller-size DB system (Figure 9a) to complete this process until 560 K. Consequently, the smaller DB system had all its quantity curves deviate from those for other systems in the ILC range from 530 K to 560 K, but it was a finite-size effect. All in all, we can conclude that heating a crystal IL can always produce the same ILC structure, and there is no need to consider the details of the initial crystal structure if we just want to have the ILC structure. It is probably the most thermodynamically stable structure in the temperature range from the melting point (up to 530 K) to the clearing point of about 655 K.

Based on our simulation results, we would like to remark on some experimental results in the literature. In the experiments of References [3–7,9–12] studying phase behaviors of ILs, the crystalline samples were measured during both heating and cooling. However, when we cooled the systems by MD simulation, they could only form SmA ILCs and dynamically go into a glassy state. Crystallization was practically almost impossible by MD simulation, let alone different crystal polymorphs, so we only presented the heating process. The comparisons between our simulation results and experimental results focused on layer spacing in the crystal phase or the ILC phase measured with X-ray scattering, as well as on the potential energy changes during phase transitions, more qualitatively, with the transition enthalpy measured with DSC. The comparison was limited to the species of [C_{14}Mim]X with a size (ionic volume) of the anion X, similar to nitrate. The electronic structure of the anion and more detailed interactions of hydrogen bonding, anion-π, etc., which may affect the IL structure, especially the structure of polar layers [30], were not considered in this work.

To our knowledge, the solid-solid phase transitions of [C_{14}Mim]X have been reported in three experimental works [6,9,12], in which the anions had similar sizes to nitrate. In Reference [6], solid-solid transitions were reported for [C_{14}Mim]$_2$[PdCl$_4$], which were irreversible for the low-temperature crystal grown from the solution and reversible for the high-temperature crystals occurring afterwards (along with the solid-LC transition). The layer spacing changed from 5.21 nm to 3.52 nm, and then to 3.72 nm for the crystal phases, and to 3.03 nm for the ILC phase. All these transitions had large transition enthalpies ranging from 16.1 to 18.8 kJ/mol, analogous to our observation of the DB structure melting into the LC. In Reference [9], solid-solid transitions were observed for [C_{14}Mim][PF$_6$], where the layer spacing had a smaller increase and the transition enthalpies were reported to be much smaller than in Reference [6]. Although the EM2 structure was modified from the structure reported in this work, the transition observed in the experiment was more like the solid-solid transition of EM1 observed in our simulation by an increase in the degree of motion in the anion and alkyl chain after the transition indicated by order parameters. In Reference [12], the solid-solid transition changed the bilayer structure to a double or extended bilayer structure for [C_{14}Mim][BF$_4$]. The layer spacing of the bilayer structure was close to HM1, EM1, or EM2, and for the double or extended bilayer was close to the DB. Such a transition was not observed in our MD simulations.

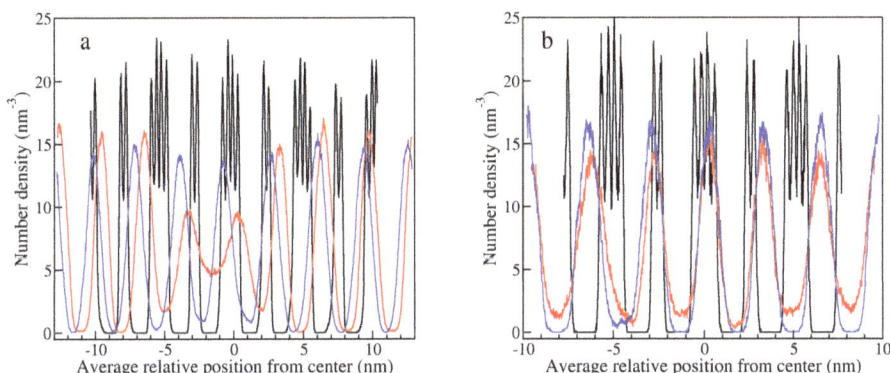

Figure 9. Average number densities of nitrates along the z axis for (**a**) the DB structure with 384 ion pairs at 530 K in the crystal phase (black), at 530 K after the solid-LC transition (red), and at 560 K in the equilibrated LC phase (blue); and (**b**) for the DB structure with 648 ion pairs, in the crystal phase (black), after the solid-LC transition when the system was not well equilibrated (red), and in the equilibrated LC phase (blue), all at 530 K.

The metastable state was in accord with the transient state reported in Reference [11] and the low-temperature smectic phase in Reference [12], as described in our previous work [14]. When the metastable state melted into the SmA LC phase, the layer spacing of the ILC computed from simulations was close to the experimental results for the [C_{14}Mim] salt with a similar-sized anion [6,7].

In experimental works [28], the DB structure has been proposed to explain the peak corresponding to spatial periods twice as large as those of bilayer structures in the direction perpendicular to the polar layers in X-ray scattering experiments, while another proposed structure has been the extended bilayer crystal structure [7,11,12], in which the alkyl chains are assumed to be in the end-to-end style, which was not observed in our MD simulations. The extended bilayer crystal structure seems unreasonable, since the area of the polar layers is incompatible with the cross-section of the nonpolar region if alkyl chains are in the end-to-end style, and thus the alkyl chains must bend and cannot result in a doubled periodicity. On the other hand, the experimental inference of heating a DB structure to ILC and then recrystallizing to a bilayer structure [7] seems to be contradictory to the fact that we have described that the DB structure is probably thermodynamically more stable than the bilayer structure at higher temperatures. A possible explanation is that the growth of the DB structure is easy in solvent [6,7], but very difficult in ILC, and hence it has never been reported. Another possibility concerns Ostwald's rule [48], according to which the least stable polymorph, here the bilayer structure, crystallizes first and becomes more stable as the temperature decreases, leading to the disappearance of DBs.

5. Conclusions

The thermotropic phase behaviors of IL [C_{14}Mim][NO_3] during the heating of five distinct crystal polymorphs were investigated by MD simulation. The simulations started from the initial crystal structures at 200 K and ended at the clearing point of ILC, about 655 K, when the systems were in the liquid phase. The phases and phase transitions in between were recognized by measuring the potential energy, densities, layer spacing of crystals or LCs, and several order parameters. By analyzing the simulation data, we could reach the following three points. First, the IL is a special enantiotropic system characterizing an entropy-favored polymorph tending to have a higher melting point. Second, there always exists a metastable state between the crystal and the SmA LC, which has the feature that the alkyl chains become perpendicular to the polar layers and have more disordered conformations. Third, the structure of the SmA LC is irrelevant to the thermal history (i.e., different crystal polymorphs melt into the same SmA structure). The first point deserves further studies and may serve as an example for soft matter that entropy is of great importance in thermal stabilities. The second point presents a common state that all crystal polymorphs must go through during the melting of crystal into SmA LC, and may help uncover the melting process of similar ILC-forming ILs. The third point reveals that the SmA LC structure is independent of the crystal structure at a lower temperature, which is practically valuable for the study of ILC by MD simulation.

Author Contributions: W.C. and Y.W. conceptualized the research, W.C. performed the simulation and analyzed the data, and W.C. and Y.W. discussed the results and wrote the paper.

Funding: This research was funded by the National Natural Science Foundation of China (Nos. 11774357 and 11747601). This work was also partly supported by the Chinese Academy of Sciences through the CAS Biophysics Interdisciplinary Innovation Team Project (No. 2060299) to Y.W.

Acknowledgments: The authors thank Giacomo Saielli for his helpful discussions. The computations of this work were conducted on the HPC cluster of ITP-CAS and the Tianhe-2 supercomputer.

Conflicts of Interest: The authors declare no conflicts of interest.

References

1. Binnemans, K. Ionic Liquid Crystals. *Chem. Rev.* **2005**, *105*, 4148–4204. [CrossRef] [PubMed]
2. Goossens, K.; Lava, K.; Bielawski, C.W.; Binnemans, K. Ionic Liquid Crystals: Versatile Materials. *Chem. Rev.* **2016**, *116*, 4643–4807. [CrossRef] [PubMed]
3. Nelyubina, Y.V.; Shaplov, A.S.; Lozinskaya, E.I.; Buzin, M.I.; Vygodskii, Y.S. A New Volume-Based Approach for Predicting Thermophysical Behavior of Ionic Liquids and Ionic Liquid Crystals. *J. Am. Chem. Soc.* **2016**, *138*, 10076–10079. [CrossRef] [PubMed]
4. Gordon, C.M.; Holbrey, J.D.; Kennedy, A.R.; Seddon, K.R. Ionic liquid crystals: Hexafluorophosphate salts. *J. Mater. Chem.* **1998**, *8*, 2627–2636. [CrossRef]
5. Holbrey, J.D.; Seddon, K.R. The phase behaviour of 1-alkyl-3-methylimidazolium tetrafluoroborates; ionic liquids and ionic liquid crystals. *J. Chem. Soc. Dalton Trans.* **1999**, 2133–2140. [CrossRef]
6. Hardacre, C.; Holbrey, J.D.; McCormac, P.B.; McMath, S.E.J.; Nieuwenhuyzen, M.; Seddon, K.R. Crystal and liquid crystalline polymorphism in 1-alkyl-3-methylimidazolium tetrachloropalladate(ii) salts. *J. Mater. Chem.* **2001**, *11*, 346–350. [CrossRef]
7. Bradley, A.E.; Hardacre, C.; Holbrey, J.D.; Johnston, S.; McMath, S.E.J.; Nieuwenhuyzen, M. Small-Angle X-ray Scattering Studies of Liquid Crystalline 1-Alkyl-3-methylimidazolium Salts. *Chem. Mater.* **2002**, *14*, 629–635. [CrossRef]

8. Masafumi, Y.; Tomohiro, M.; Kiyoshi, K.; Masahiro, Y.; Hiroyuki, O.; Takashi, K. Liquid-Crystalline Assemblies Containing Ionic Liquids: An Approach to Anisotropic Ionic Materials. *Chem. Lett.* **2002**, *31*, 320–321. [CrossRef]

9. De Roche, J.; Gordon, C.M.; Imrie, C.T.; Ingram, M.D.; Kennedy, A.R.; Lo Celso, F.; Triolo, A. Application of complementary experimental techniques to characterization of the phase behavior of [C16mim][PF6] and [C14mim][PF6]. *Chem. Mater.* **2003**, *15*, 3089–3097. [CrossRef]

10. Guillet, E.; Imbert, D.; Scopelliti, R.; Bünzli, J.-C.G. Tuning the Emission Color of Europium-Containing Ionic Liquid-Crystalline Phases. *Chem. Mater.* **2004**, *16*, 4063–4070. [CrossRef]

11. Li, L.; Groenewold, J.; Picken, S.J. Transient Phase-Induced Nucleation in Ionic Liquid Crystals and Size-Frustrated Thickening. *Chem. Mater.* **2005**, *17*, 250–257. [CrossRef]

12. Nozaki, Y.; Yamaguchi, K.; Tomida, K.; Taniguchi, N.; Hara, H.; Takikawa, Y.; Sadakane, K.; Nakamura, K.; Konishi, T.; Fukao, K. Phase Transition and Dynamics in Imidazolium-Based Ionic Liquid Crystals through a Metastable Highly Ordered Smectic Phase. *J. Phys. Chem. B* **2016**, *120*, 5291–5300. [CrossRef] [PubMed]

13. Saielli, G. MD simulation of the mesomorphic behaviour of 1-hexadecyl-3-methylimidazolium nitrate: Assessment of the performance of a coarse-grained force field. *Soft Matter* **2012**, *8*, 10279–10287. [CrossRef]

14. Cao, W.; Wang, Y.; Saielli, G. Metastable State during Melting and Solid–Solid Phase Transition of [CnMim][NO3] (n = 4–12) Ionic Liquids by Molecular Dynamics Simulation. *J. Phys. Chem. B* **2018**, *122*, 229–239. [CrossRef] [PubMed]

15. Quevillon, M.J.; Whitmer, J.K. Charge Transport and Phase Behavior of Imidazolium-Based Ionic Liquid Crystals from Fully Atomistic Simulations. *Materials* **2018**, *11*. [CrossRef]

16. Peng, H.; Kubo, M.; Shiba, H. Molecular dynamics study of mesophase transitions upon annealing of imidazolium-based ionic liquids with long-alkyl chains. *Phys. Chem. Chem. Phys.* **2018**, *20*, 9796–9805. [CrossRef]

17. Saielli, G.; Bagno, A.; Wang, Y. Insights on the Isotropic-to-Smectic A Transition in Ionic Liquid Crystals from Coarse-Grained Molecular Dynamics Simulations: The Role of Microphase Segregation. *J. Phys. Chem. B* **2015**, *119*, 3829–3836. [CrossRef] [PubMed]

18. Jayaraman, S.; Maginn, E.J. Computing the melting point and thermodynamic stability of the orthorhombic and monoclinic crystalline polymorphs of the ionic liquid 1-n-butyl-3-methylimidazolium chloride. *J. Chem. Phys.* **2007**, *127*, 214504. [CrossRef] [PubMed]

19. Reichert, W.M.; Holbrey, J.D.; Swatloski, R.P.; Gutowski, K.E.; Visser, A.E.; Nieuwenhuyzen, M.; Seddon, K.R.; Rogers, R.D. Solid-State Analysis of Low-Melting 1,3-Dialkylimidazolium Hexafluorophosphate Salts (Ionic Liquids) by Combined X-ray Crystallographic and Computational Analyses. *Cryst. Growth Des.* **2007**, *7*, 1106–1114. [CrossRef]

20. Saielli, G.; Voth, G.A.; Wang, Y. Diffusion mechanisms in smectic ionic liquid crystals: Insights from coarse-grained MD simulations. *Soft Matter* **2013**, *9*, 5716–5725. [CrossRef]

21. Saielli, G.; Wang, Y. Role of the Electrostatic Interactions in the Stabilization of Ionic Liquid Crystals: Insights from Coarse-Grained MD Simulations of an Imidazolium Model. *J. Phys. Chem. B* **2016**, *120*, 9152–9160. [CrossRef] [PubMed]

22. Saielli, G. Fully Atomistic Simulations of the Ionic Liquid Crystal [C16mim][NO3]: Orientational Order Parameters and Voids Distribution. *J. Phys. Chem. B* **2016**, *120*, 2569–2577. [CrossRef] [PubMed]

23. Ji, Y.; Shi, R.; Wang, Y.; Saielli, G. Effect of the Chain Length on the Structure of Ionic Liquids: From Spatial Heterogeneity to Ionic Liquid Crystals. *J. Phys. Chem. B* **2013**, *117*, 1104–1109. [CrossRef] [PubMed]

24. Bruce, A.D.; Wilding, N.B.; Ackland, G.J. Free Energy of Crystalline Solids: A Lattice-Switch Monte Carlo Method. *Phys. Rev. Lett.* **1997**, *79*, 3002–3005. [CrossRef]

25. Bruce, A.D.; Jackson, A.N.; Ackland, G.J.; Wilding, N.B. Lattice-switch Monte Carlo method. *Phys. Rev. E* **2000**, *61*, 906–919. [CrossRef]

26. Wilding, N.B.; Bruce, A.D. Freezing by Monte Carlo Phase Switch. *Phys. Rev. Lett.* **2000**, *85*, 5138–5141. [CrossRef]

27. Jackson, A.N.; Bruce, A.D.; Ackland, G.J. Lattice-switch Monte Carlo method: Application to soft potentials. *Phys. Rev. E* **2002**, *65*, 036710. [CrossRef]

28. Downard, A.; Earle, M.J.; Hardacre, C.; McMath, S.E.J.; Nieuwenhuyzen, M.; Teat, S.J. Structural Studies of Crystalline 1-Alkyl-3-Methylimidazolium Chloride Salts. *Chem. Mater.* **2004**, *16*, 43–48. [CrossRef]

29. Greaves, T.L.; Kennedy, D.F.; Mudie, S.T.; Drummond, C.J. Diversity Observed in the Nanostructure of Protic Ionic Liquids. *J. Phys. Chem. B* **2010**, *114*, 10022–10031. [CrossRef]
30. Richard, P.M.; Claire, A.; Tom, W.; Patricia, A.H. The impact of anion electronic structure: Similarities and differences in imidazolium based ionic liquids. *J. Phys. Condens. Matter* **2014**, *26*, 284112. [CrossRef]
31. Wang, Y.; Voth, G.A. Unique Spatial Heterogeneity in Ionic Liquids. *J. Am. Chem. Soc.* **2005**, *127*, 12192–12193. [CrossRef] [PubMed]
32. Wang, Y.; Voth, G.A. Tail Aggregation and Domain Diffusion in Ionic Liquids. *J. Phys. Chem. B* **2006**, *110*, 18601–18608. [CrossRef] [PubMed]
33. Canongia Lopes, J.N.A.; Pádua, A.A.H. Nanostructural Organization in Ionic Liquids. *J. Phys. Chem. B* **2006**, *110*, 3330–3335. [CrossRef] [PubMed]
34. Cornell, W.D.; Cieplak, P.; Bayly, C.I.; Gould, I.R.; Merz, K.M.; Ferguson, D.M.; Spellmeyer, D.C.; Fox, T.; Caldwell, J.W.; Kollman, P.A. A Second Generation Force Field for the Simulation of Proteins, Nucleic Acids, and Organic Molecules. *J. Am. Chem. Soc.* **1995**, *117*, 5179–5197. [CrossRef]
35. Frisch, M.J.; Trucks, G.W.; Schlegel, H.B.; Scuseria, G.E.; Robb, M.A.; Cheeseman, J.R.; Scalmani, G.; Barone, V.; Mennucci, B.; Petersson, G.A.; Revision, E.; et al. *Gaussian 09*; Revision E.01; Gaussian, Inc.: Wallingford, CT, USA, 2013.
36. Singh, U.C.; Kollman, P.A. An approach to computing electrostatic charges for molecules. *J. Comput. Chem.* **1984**, *5*, 129–145. [CrossRef]
37. Besler, B.H.; Merz, K.M.; Kollman, P.A. Atomic charges derived from semiempirical methods. *J. Comput. Chem.* **1990**, *11*, 431–439. [CrossRef]
38. Bayly, C.I.; Cieplak, P.; Cornell, W.; Kollman, P.A. A well-behaved electrostatic potential based method using charge restraints for deriving atomic charges: The RESP model. *J. Phys. Chem.* **1993**, *97*, 10269–10280. [CrossRef]
39. Abraham, M.J.; Murtola, T.; Schulz, R.; Páll, S.; Smith, J.C.; Hess, B.; Lindahl, E. GROMACS: High performance molecular simulations through multi-level parallelism from laptops to supercomputers. *SoftwareX* **2015**, *1–2*, 19–25. [CrossRef]
40. Parrinello, M.; Rahman, A. Polymorphic transitions in single crystals: A new molecular dynamics method. *J. Appl. Phys.* **1981**, *52*, 7182–7190. [CrossRef]
41. Nosé, S. A molecular dynamics method for simulations in the canonical ensemble. *Mol. Phys.* **1984**, *52*, 255–268. [CrossRef]
42. Darden, T.; York, D.; Pedersen, L. Particle mesh Ewald: An N·log(N) method for Ewald sums in large systems. *J. Chem. Phys.* **1993**, *98*, 10089–10092. [CrossRef]
43. Humphrey, W.; Dalke, A.; Schulten, K. VMD: Visual molecular dynamics. *J. Mol. Graph.* **1996**, *14*, 33–38. [CrossRef]
44. Bates, M.A.; Luckhurst, G.R. Computer simulation studies of anisotropic systems. XXX. The phase behavior and structure of a Gay-Berne mesogen. *J. Chem. Phys.* **1999**, *110*, 7087–7108. [CrossRef]
45. Feng, H.; Zhou, J.; Qian, Y. Atomistic simulations of the solid-liquid transition of 1-ethyl-3-methyl imidazolium bromide ionic liquid. *J. Chem. Phys.* **2011**, *135*, 144501. [CrossRef]
46. Zhang, Y.; Maginn, E.J. A comparison of methods for melting point calculation using molecular dynamics simulations. *J. Chem. Phys.* **2012**, *136*, 144116. [CrossRef] [PubMed]
47. Zhang, Y.; Maginn, E.J. Molecular dynamics study of the effect of alkyl chain length on melting points of [CnMIM][PF6] ionic liquids. *Phys. Chem. Chem. Phys.* **2014**, *16*, 13489–13499. [CrossRef] [PubMed]
48. Ostwald, W. Studien über die Bildung und Umwandlung fester Körper. *Z. Phys. Chem.* **1897**, *22U*, 289. [CrossRef]

![crystals logo] *crystals*

MDPI

Article

Molecular and Segmental Orientational Order in a Smectic Mesophase of a Thermotropic Ionic Liquid Crystal

Jing Dai [1], Boris B. Kharkov [2] and Sergey V. Dvinskikh [1,2,*]

[1] Department of Chemistry, KTH Royal Institute of Technology, SE-10044 Stockholm, Sweden; jdai@kth.se
[2] Laboratory of Biomolecular NMR, Saint Petersburg State University, Saint Petersburg 199034, Russia; st036820@student.spbu.ru
*   Correspondence: sergeid@kth.se; Tel.: +46-8-790-82-24

Received: 30 November 2018; Accepted: 24 December 2018; Published: 28 December 2018

Abstract: We investigate conformational dynamics in the smectic A phase formed by the mesogenic ionic liquid 1-tetradecyl-3-methylimidazolium nitrate. Solid-state high-resolution ^{13}C nuclear magnetic resonance (NMR) spectra are recorded in the sample with the mesophase director aligned in the magnetic field of the NMR spectrometer. The applied NMR method, proton encoded local field spectroscopy, delivers heteronuclear dipolar couplings of each ^{13}C spin to its ^{1}H neighbours. From the analysis of the dipolar couplings, orientational order parameters of the C–H bonds along the hydrocarbon chain were determined. The estimated value of the molecular order parameter S is significantly lower compared to that in smectic phases of conventional non-ionic liquid crystals.

Keywords: ionic liquids; liquid crystals; ionic liquid crystals; molecular orientational order; nuclear magnetic resonance

1. Introduction

Ionic liquids that form ordered mesophases in temperature ranges between solid and isotropic liquid phases belong to a class of ionic liquid crystal (ILC) materials [1]. Molecules and ions in liquid crystals (LC) exhibit a high molecular translational and rotational mobility combined with orientational and positional order. The unique synergy of ionic conductivity and anisotropic nanoscale structure makes ILC materials in high demand in the development of low-dimensional charge transport technology and other applications [2].

Nuclear magnetic resonance (NMR) spectroscopy is widely used to study molecular conformational and rotational dynamics and orientational order in LC [3,4]. Site-specific information at the atomic level can be obtained by high resolution ^{13}C NMR combined with selective suppression of spin interactions. Structural and order parameters are obtained through the measurements of orientation- and distance-dependent dipole couplings. Due to the orientational ordering, the dipolar coupling in mesophase is not averaged to zero, in contrast to isotropic phase. ^{13}C–^{1}H heteronuclear dipolar couplings in LC are measured by two-dimensional (2D) separated local field spectroscopy technique [5].

The local bond order parameter S_{CH} characterizes the average orientation of the C–H bond to the LC director. The ratio of the residual dipolar coupling averaged by the anisotropic motion, to its unaveraged value provides the S_{CH}. For molecules with long hydrocarbon chains one can study a C–H bond order parameter profile, a variation of S_{CH} along the chain, which characterizes the average molecular structure in the mesophase [6–8].

Here we investigate conformation and molecular order of organic cations in the smectic A phase of the mesogenic ionic liquid 1-tetradecyl-3-methylimidazolium nitrate (C_{14}mimNO$_3$) [9].

The sample forms interdigitated bilayer structures where the molecules, consisting of non-ionic and ionic non-miscible parts, phase-separate at the nanoscale. Relatively low transition temperatures between meso- and isotropic phases and a wide temperature range of the mesophase existence make this sample particularly suitable for preparing it both in un-aligned state and with director aligned with respect to external magnetic field. In the latter case, the sample is suitable for studies by high-resolution solid-state NMR at stationary condition (without sample spinning). We apply 2D ^{13}C–^{1}H dipolar NMR spectroscopy to the Smectic A phase of the C$_{14}$mimNO$_3$ to quantitatively characterize the molecular dynamics. Based on the profiles of the order parameters obtained, we put forward a model for the motion of the organic cations in the bilayer of the smectic A phase.

2. Materials and Methods

The sample of C$_{14}$mimNO$_3$ (1-tetradecyl-3-methylimidazolium nitrate, CAS 799246-94-9) was obtained from Angene International and used as received. Water content in the sample was estimated to ≈1.0 wt.% from ^{1}H NMR spectrum recorded in the isotropic phase. No other significant impurity signals were observed. The sample exhibited the following phase transition temperatures: Isotropic $\xrightarrow{+129\,°C}$ Smectic A $\xrightarrow{+43\,°C}$ Crystal, as observed by recording proton NMR spectra. Transition temperature to the smectic phase is in agreement with reported phase diagram [9]. Slightly higher crystallization temperature compared to that determined by differential scanning calorimetry in [9] is presumably due to much slower cooling rate applied in our NMR experiments. Upon slow cooling from the isotropic phase in the presence of the strong external magnetic field, director of the smectic phase is aligned perpendicular to the magnetic field vector. Most of our experiments were performed in such an oriented sample. To produce random director orientation in the smectic phase, the sample was cooled from the isotropic phase while out of the magnet. Due to the high viscosity of the smectic phase, the domains of different orientations do not reorient when the sample in the smectic phase is placed in magnetic field.

Experiments were performed using Bruker 500 Avance III spectrometer at Larmor frequencies of 500.1 and 125.7 for ^{1}H and ^{13}C, respectively. About 0.5 g of the sample was loaded in a standard 5 mm NMR tube. NMR spectra were recorded using solution state multinuclear 5 mm probe-head. The ^{1}H and ^{13}C 90°-pulse lengths were 8 and 13 μs, respectively. For heteronuclear proton decoupling in the mesophase, Spinal64 sequence [10] with the ^{1}H nutation frequency of 23 kHz was used during acquisition time of 120 ms. To enhance the intensity of the ^{13}C signal, proton-to-carbon cross polarization (CP) with adiabatic demagnetization in the rotating frame (ADRF) [11] was applied with nutation frequencies up to 16 kHz and contact time in the range 10–20 ms.

Dipolar ^{1}H–^{13}C spectra were recorded using proton detected/encoded local field (PDLF) NMR spectroscopy [12]. The PDLF spectrum is governed by a two-spin interaction and thus cross sections in the indirect dimension present superposition of dipolar doublets. The actual PDLF pulse sequence is shown in the Supplementary Materials, Figure S2. The evolution time in indirect time domain was incremented with 384 μs in 256 steps, at each with four collected transients. Proton homonuclear decoupling during the evolution time was achieved by the BLEW-48 multiple-pulse sequence [13] with a nutation frequency of 31.2 kHz.

The temperature was regulated with an accuracy of 0.1 °C. The temperature shift and temperature gradient within the sample, caused by the decoupling irradiation, were calibrated by observing the change in the ^{13}C spectral line widths and positions. Decoupling power, irradiation time, and repetition delay were adjusted to limit heating effects to <0.5 °C.

3. Results and Discussion

3.1. Carbon-13 Nuclear Magnetic Resonance (NMR) Spectra

Representative ^{13}C NMR spectra of the C_{14}mim ions in different phases are displayed in Figure 1. All carbons, except carbons 5 and 6 with overlapped signals, were resolved. The assignment of the carbon signals in the isotropic and mesophase was verified, respectively, by the INADEQUATE (incredible natural-abundance double-quantum transfer) experiment [14] and by dipolar INADEQUATE experiment as described in Supplementary Materials. The chemical shifts of the corresponding signals in the spectra in Figure 1a,b reflect different parts of residual chemical shift anisotropy (CSA) tensor. The signal positions in the isotropic phase are given by the isotropic averages δ_i of the respective CSA tensors. In the spectrum in the aligned mesophase, (Figure 1b), the chemical shift corresponds to one of the principal values $\delta_{\alpha\alpha}$ ($\alpha = x, y, z$) of the residual CSA tensor.

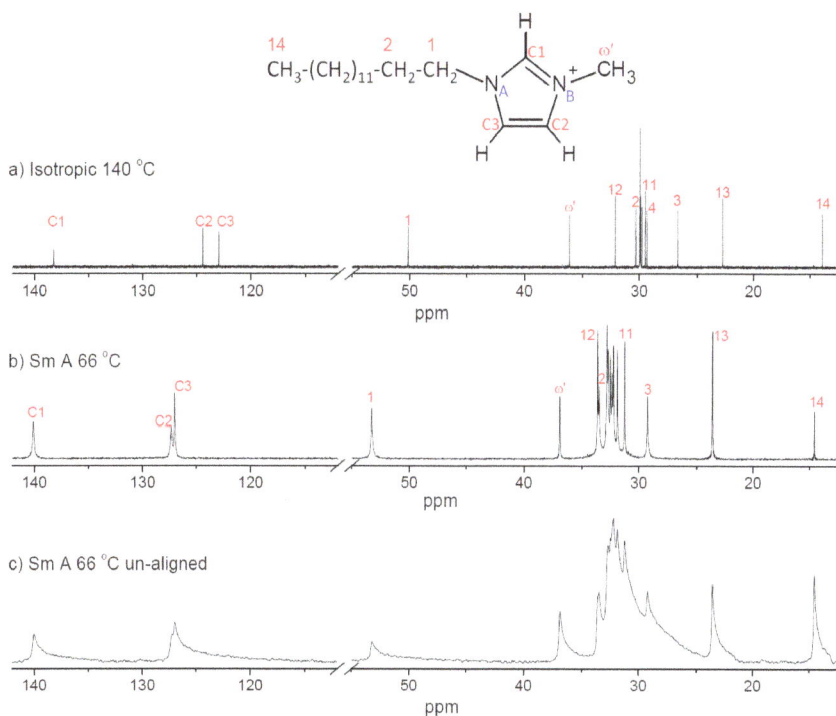

Figure 1. ^{13}C nuclear magnetic resonance (NMR) spectra of C_{14}mimNO$_3$ in the isotropic (**a**) and smectic A phase (**b,c**) at indicated temperatures. The spectrum (**c**) was recorded in the un-aligned smectic A phase.

In uniaxial liquid crystals, the chemical shift tensor is described by the components along and orthogonal to the phase director, $\delta_{||}^{LC}$ and $\delta_{\perp}^{LC}$, respectively. In our sample, which exhibits a negative anisotropy of the diamagnetic susceptibility, the director aligns in the plane perpendicular to the magnetic field of the spectrometer. Hence, the observed chemical shifts are determined by the $\delta_{\perp}^{LC}$ values. This is verified comparing the spectrum of the aligned sample in Figure 1b to the spectrum of the sample prepared with random director distribution, Figure 1c. The random director alignment was achieved by cooling the sample from the isotropic phase outside the NMR magnet. The chemical shifts corresponding to the edges of the axially symmetric CSA patterns observed in un-aligned sample, match those of the respective lines in the aligned mesophase.

3.2. ^{1}H–^{13}C Dipolar Spectra

The 2D PDLF spectrum at 66 °C is shown in Figure 2a. The cross sections in the dipolar direction for different carbons are shown in Figure 2b. The dipolar splittings between directly bound ^{13}C and ^{1}H spins are well resolved and can be directly measured in the spectrum for each carbon in the molecule. Additional low intensity doublets observed for some of the dipolar cross-sections, e.g., in the spectrum of the chain carbon 2, are due to partial overlap with nearby signals in the 2D spectrum. The inner multiplets with small splittings resolved for some carbon sites arise from couplings to remote protons in neighbouring groups.

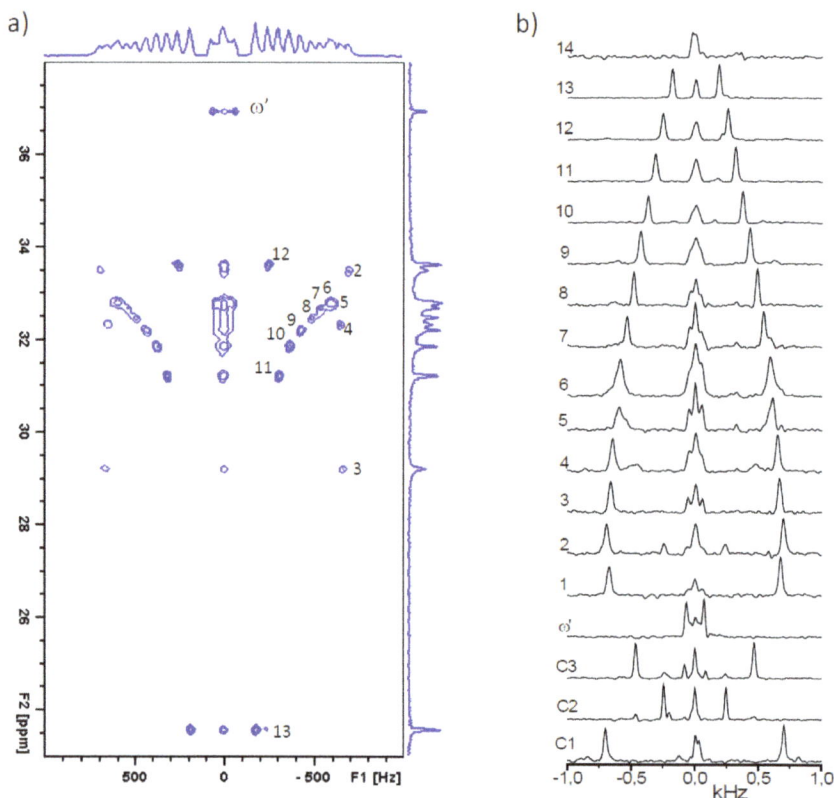

Figure 2. (a) Part of 2D proton detected/encoded local field (PDLF) spectrum in C_{14}mimNO$_3$ smectic A phase at 66 °C. (b) Cross-sections along dipolar dimension are shown for all carbons. For spectral assignment, see molecular structure in Figure 1.

The dipole–dipole spin couplings are orientation- and interatomic distances dependent. Anisotropic molecular motion results in partial averaging of the intramolecular dipolar interactions over the conformational dynamics, molecular reorientations, and the molecular axis fluctuations about the mesophase director. Intermolecular spin couplings are averaged to zero by translational dynamics [15]. Different motional modes can often be considered uncorrelated and their effects on spin interactions can be separated. A simplified description of orientational ordering can be applied considering an approximately cylindrical symmetry due to fast uniaxial rotations of elongated mesogenic molecules [16]. For a ^1H–^{13}C spin pair in a molecule in an anisotropic phase the dipolar coupling is:

$$d_{CH} = b_{CH}\langle P_2(\cos\theta_{PL})\rangle \tag{1}$$

with $P_2(\cos\theta_{PL}) = (3\cos^2\theta_{PL} - 1)/2$, where θ_{PL} is the angle between the C–H vector, which defines a principal frame axis P of the dipolar interaction, and the magnetic field B_0 (defines the laboratory frame axis L). The dipolar coupling constant in the principal frame $b_{CH} = -(\mu_0/8\pi^2)(\gamma_H\gamma_C\hbar/r_{CH}^3)$ can be estimated from the atomic distances r_{CH} and the gyromagnetic ratios γ_H, γ_C. For a single C–H bond with account for vibration effects we accept b_{CH} values of −21.5 kHz and −22 kHz for aliphatic and aromatic sites, respectively [17,18].

Anisotropic molecular mobility leads to a partial averaging of the angular term $\langle P_2(\cos\theta_{PL})\rangle$. Dynamic molecular disorder in LC is measured by the molecular orientational order parameter $S = \langle P_2(\cos\theta_{MN})\rangle$, where θ_{MN} is the instantaneous angle between the long molecular axis M and the director N. Similarly, the orientational averaging of the local C–H bond direction with the respect to director is described by the bond order parameter $S_{CH} = \langle P_2(\cos\theta_{PN})\rangle$, where θ_{PN} is the angle between the inter-nuclear vector P and the director N. Thus,

$$d_{CH} = b_{CH}S_{CH}P_2(\cos\theta_{NL}) \tag{2}$$

where θ_{NL} is the angle between director and magnetic field vector. Hence, local bond order parameters S_{CH} can be estimated directly from the NMR dipolar spectra. Assuming statistical independence and large time-scale separation of molecular reorientation and conformational dynamics, the bond order parameters S_{CH} can be related to the molecular order parameter S by the expression:

$$S_{CH} = \langle P_2(\cos\theta_{PM})\rangle S \tag{3}$$

where θ_{PM} defines the angle between the bond vector and molecular axis. Thus, Equation (2) can be expanded:

$$d_{CII} - b_{CII}\langle P_2(\cos\theta_{PM})\rangle SP_2(\cos\theta_{NL}) \tag{4}$$

For non-rigid molecules, the separation of the terms $\langle P_2(\cos\theta_{PM})\rangle$ and S requires some model assumptions of molecular conformational dynamics.

The splitting $\Delta\nu$ observed in the PDLF experiment is contributed by residual dipolar coupling d_{CH} and isotropic indirect spin coupling J_{CH}:

$$\Delta\nu = k(2d_{CH} + J_{CH}) \tag{5}$$

(anisotropic part of indirect coupling ΔJ_{CH} is neglected compared with dipolar coupling [16]). Upon application of homonuclear decoupling both these interactions are scaled down. For the pulse sequence BLEW48 [13], the scaling factor was experimentally calibrated to $k = 0.418 \pm 0.02$.

The magnitudes of the isotropic coupling J_{CH} were measured in the isotropic phase by recording ^{13}C NMR spectrum without proton decoupling. The sign of J_{CH} is known to be positive. Note that the sign of the splitting $\Delta\nu$ is not obtained in the PDLF experiment. However, the sign of the dipolar coupling d_{CH} can be determined from Equation (4) considering average orientation of molecular axis in the magnetic field and C–H bond angles with respect to molecular axis. The dipolar coupling constant in the principal frame b_{CH} is negative for a proton-carbon pair. The molecular order parameter S is positive. The term $P_2(\cos\theta_{NL}) = -0.5$ for the director aligned perpendicular to the magnetic field. The sign of the angular term $P_2(\cos\theta_{PM})$ can be deduced by considering molecular geometry as discussed below.

3.3. Order Parameters

With dipolar couplings d_{CH} obtained from Equation (5), local bond order parameters S_{CH} are directly calculated using Equation (2). With some assumptions about the direction of the molecular axis, as discussed below, molecular order parameter S can as well be estimated from Equation (3). Thus, the analysis on the PDLF spectra provides quantitative information on the local and molecular order parameters, S_{CH} and S.

The variations of S_{CH} along the alkyl chain, or order parameter profiles, are plotted in Figure 3 for selected temperatures in the smectic A phase. The S_{CH} values are negative assuming the average orientation of the C–H bond perpendicular to the direction of long molecular axis. As expected, the methylene groups close to the imidazolium core are less disordered and segmental mobility increases gradually towards the chain terminal methyl. While this average character of the chain dynamics persists in the whole temperature range of the smectic phase, the alkyl chains become more ordered at lower temperatures.

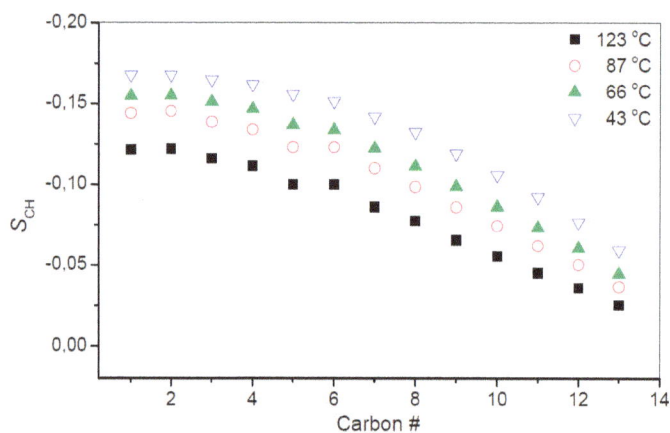

Figure 3. C–H bond order parameter S_{CH} profiles for the alkyl chain of the cation in the smectic A phase of $C_{14}mimNO_3$.

The S_{CH} magnitudes are small and similar to those typically found in lyotropic lamellar phases [8]. It is interesting that S_{CH} values obtained by MD simulation for $C_{16}mimNO_3$ homologue were also found to be similarly small [19]. The low values of S_{CH}, on one hand, may indicate a significant conformational dynamics by trans-gauche isomerisation. The fast decrease of S_{CH} towards chain terminal suggests a gradual intensification of the conformational mobility with increasingly significant population of gauche conformers. On the other hand, limited variation of the order parameters for the first few methylene groups close to the imidazolium core suggests a restricted conformation dynamics for this part of the chain with predominant trans conformation. In this case, relatively small observed values of S_{CH} parameters for the carbons in vicinity of the head group can be due to either a low value of the molecular order parameter S (cf. Equation (3)) or a significant tilt angle of this part of the chain with respect to main molecular axis oriented along the director. Based on this observation, we consider the following model of average molecular conformation in the further analysis: the part of the chain in vicinity of the head group is predominantly in the trans conformation and the symmetry axis of this chain fragment does not significantly deviate from the long molecular axis. The latter assumption is, in fact, supported by the values of S_{CC} order parameters, estimated from the $^{13}C_n-^{13}C_{n+2}$ dipolar couplings for the chain carbons separated by two bonds. The coupling constants were obtained from the dipolar INADEQUATE experiment described in the Supplementary Materials. The estimated magnitude of the order parameters $S_{CC} \approx 0.31$ at 66 °C (averaged value over methylene segments 1 to 5) is a factor of two larger than the corresponding value $S_{CH} \approx -0.16$. The ratio $S_{CC}/S_{CH} \approx -2$ is indeed expected for C_n–C_{n+2} and C–H vectors, respectively, along and perpendicular to the molecular axis, and thus the model assumption is corroborated.

Within this approach, for the methylenes close to the head group the angular term is $P_2(\cos\theta_{PM}) \approx -0.5$. The molecular order parameter S is, thus, estimated using Equation (4). Temperature dependence $S(T)$ is displayed in Figure 4b, while temperature dependencies of the local bond order parameters for the imidazolium core and for the selected carbons in the chain are shown in Figure 4a. We conclude that the ionic smectic A phase is formed (counter-intuitively) with significantly lower value of the molecular order parameter S when comparing to non-ionic smectic mesogens [20–22]. Theoretical and computational studies suggest that, in contrast to neutral mesogens, the orientational ordering is less important for the stabilization of the smectic layers in ILC where dominant stabilizing effect is due to a "charge-ordered" nanoscale segregation induced by ionic interactions [1,19,23,24].

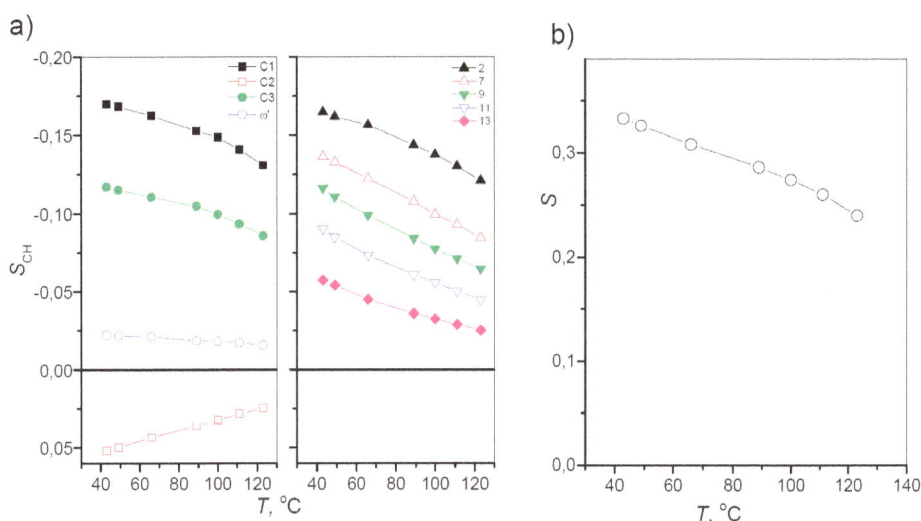

Figure 4. (**a**) Temperature dependencies of the local bond order parameters S_{CH} for the imidazolium core and selected carbons in the chain in the smectic A phase of C_{14}mimNO$_3$. (**b**) Temperature dependence of the molecular order parameter S.

In order to analyse the dipolar couplings and local bond order parameters in the imidazolium core (Figure 4a), the molecular axis alignment to imidazolium ring plane has to be correctly accounted for. The planar molecular structure as displayed in Figure 1 is not consistent with the observed couplings in the imidazolium core. In fact, it has been shown by density functional theory (DFT) analysis of similar compounds that in the energy optimized equilibrium structure of $C_n$mim cations there is a significant angle between the core plane and the symmetry axis of the all-trans aliphatic chain [25]. Non-planar molecular structure is also consistent with recent molecular dynamics (MD) simulation data [19]. A rotation/flip of the imidazolium core around N–C bond can also be accompanied by the chain re-alignment. The correlation of motion by ion-pairing, as has been observed in isotropic phase of ionic liquids, may impose additional restriction on the core dynamics [26–29]. On the other hand, diffusion and conductivity studies have indicated (partial) dissociation of cations and anions in mesophases [28–30]. With reservations for the complexity of the processes controlling the core alignment and dynamics, it is still possible make rough estimates of the order parameters using DFT structure analysis as presented in [19]. Application of Equation (3) using the direction angles given in [19] results in S_{CH} estimates of −0.14, +0.04, and −0.07 for the carbons C1, C2, and C3, respectively, (at 66 °C) in rough agreement with our experimental results (Figure 4a).

4. Conclusions

In the presented work, the molecular and conformational dynamics of flexible organic cations in the smectic A phase of the ionic liquid C_{14}mimNO_3 was studied by means of ^{1}H–^{13}C dipolar NMR. We demonstrated that the solid-state dipolar NMR spectroscopy in ILC samples is a powerful tool sensitive to details of the molecular conformation and dynamics in the organic smectic layer. The approach does not require isotope labelling; the measurements are performed in samples with natural isotopic abundance. With the sample director aligned in the magnetic field of the spectrometer, the solid-state high-resolution NMR spectra can be obtained in a stationary sample without the necessity of using a magic angle spinning technique.

The obtained experimental data are consistent with the model of the chains preferentially aligned with the layer normal and characterized by relatively low values of the orientational order parameters. Within the temperature interval of the smectic phase, the molecular order parameter S is in the range of 0.25–0.35, which is in approximate agreement with the results of MD analysis of the C16-homologue sample [19,31]. The high rotational/conformational mobility of the organic component in the smectic phase is also accompanied by fast translational displacement as has been shown in experimental and computational diffusion studies of structurally similar ILC [30,32]. Thus, this and previous studies prove highly dynamic behaviour of organic cations in smectic bilayers in ILC.

Supplementary Materials: The following are available online at http://www.mdpi.com/2073-4352/9/1/18/s1, Figure S1: Temperature dependent ^{13}C NMR spectra of C_{14}mimNO_3 in isotropic and smectic A phase. Figure S2: PDLF pulse sequence to record dipolar ^{13}C–^{1}H spectra. Figure S3: Dipolar INADEQUATE spectrum in smectic A phase of C_{14}mimNO_3. Table S1: ^{13}C–^{13}C dipolar splittings and dipolar couplings for alkyl chain carbons separated by two bonds.

Author Contributions: S.V.D. designed and proposed the methods. J.D. and S.V.D. performed the NMR measurements. B.B.K. performed the numerical analysis. All authors contributed to the preparation of the manuscript.

Funding: This work was supported by the Swedish Research Council VR and by the Russian Foundation for Basic Research (project no. 17-03-00057).

Conflicts of Interest: The authors declare no conflict of interest.

References

1. Goossens, K.; Lava, K.; Bielawski, C.W.; Binnemans, K. Ionic liquid crystals: Versatile materials. *Chem. Rev.* **2016**, *116*, 4643–4807. [CrossRef] [PubMed]
2. Kato, T.; Yoshio, M.; Ichikawa, T.; Soberats, B.; Ohno, H.; Funahashi, M. Transport of ions and electrons in nanostructured liquid crystals. *Nat. Rev. Mater.* **2017**, *2*, 17001. [CrossRef]
3. Dong, R.Y. *Nuclear Magnetic Resonance Spectroscopy of Liquid Crystals*; Worlds Scientific: London, UK, 2010.
4. Dvinskikh, S.V.; Sandström, D.; Zimmermann, H.; Maliniak, A. Carbon-13 NMR Spectroscopy applied to columnar liquid crystals. *Progr. Nucl. Magn. Reson. Spectrosc.* **2006**, *48*, 85–107. [CrossRef]
5. Dvinskikh, S.V. Characterization of Liquid-Crystalline Materials by Separated Local Field Methods. In *Modern Methods in Solid-State NMR: A practitioners' Guide*; Hodgkinson, P., Ed.; Royal Society of Chemistry: Abingdon, UK, 2018.
6. Boden, N.; Clark, L.D.; Bushby, R.J.; Emsley, J.W.; Luckhurst, G.R.; Stockley, C.P. A deuterium N.M.R. study of chain ordering in the liquid crystals 4,4'-di-n-heptyloxyazoxybenzene and 4-n-octyl-4'-cyanobiphenyl. *Mol. Phys.* **1981**, *42*, 565–594. [CrossRef]
7. Dvinskikh, S.V.; Castro, V.; Sandström, D. Probing segmental order in lipid bilayers at variable hydration levels by amplitude- and phase-modulated cross-polarization NMR. *Phys. Chem. Chem. Phys.* **2005**, *7*, 3255–3257. [CrossRef] [PubMed]
8. Kharkov, B.B.; Chizhik, V.I.; Dvinskikh, S.V. Low RF power high resolution ^{1}H–^{13}C–^{14}N Separated local field spectroscopy in lyotropic mesophases. *J. Magn. Reson.* **2012**, *223*, 73–79. [CrossRef] [PubMed]

9. Guillet, E.; Imbert, D.; Scopelliti, R.; Bunzli, J.C.G. Tuning the emission color of europium-containing ionic liquid-crystalline phases. *Chem. Mater.* **2004**, *16*, 4063–4070. [CrossRef]
10. Fung, B.M.; Khitrin, A.K.; Ermolaev, K. An improved broadband decoupling sequence for liquid crystals and solids. *J. Magn. Reson.* **2000**, *142*, 97–101. [CrossRef]
11. Lee, J.S.; Khitrin, A.K. Thermodynamics of adiabatic cross polarization. *J. Chem. Phys.* **2008**, *128*, 114504. [CrossRef]
12. Dvinskikh, S.V.; Zimmermann, H.; Maliniak, A.; Sandström, D. Separated local field spectroscopy of columnar and nematic liquid crystals. *J. Magn. Reson.* **2003**, *163*, 46–55. [CrossRef]
13. Burum, D.P.; Linder, M.; Ernst, R.R. Low-power multipulse line narrowing in solid-state NMR. *J. Magn. Reson.* **1981**, *44*, 173–188. [CrossRef]
14. Berger, S.; Braun, S. *200 and More NMR Experiments: A Practical Course*; Wiley: Leipzig, Germany, 2004.
15. Dvinskikh, S.V.; Furó, I. Nuclear magnetic resonance studies of translational diffusion in thermotropic liquid crystals. *Russ. Chem. Rev.* **2006**, *75*, 497–506. [CrossRef]
16. Fung, B.M. Liquid crystalline samples: Carbon-13 NMR. In *Encyclopedia of Nuclear Magnetic Resonance*; Grant, D.M., Harris, R.K., Eds.; Wiley: Chichester, UK, 1996; pp. 2744–2751.
17. Dvinskikh, S.V.; Zimmermann, H.; Maliniak, A.; Sandström, D. Measurements of motionally averaged heteronuclear dipolar couplings in MAS NMR using R-type recoupling. *J. Magn. Reson.* **2004**, *168*, 194–201. [CrossRef] [PubMed]
18. Dvinskikh, S.V.; Sandström, D. Frequency offset refocused PISEMA-type sequences. *J. Magn. Reson.* **2005**, *175*, 163–169. [CrossRef] [PubMed]
19. Saielli, G. Fully Atomistic Simulations of the ionic liquid crystal [C(16)mim][NO$_3$]: Orientational order parameters and voids distribution. *J. Phys. Chem. B* **2016**, *120*, 2569–2577. [CrossRef] [PubMed]
20. Constant, M.; Decoster, D. Raman-scattering—Investigation of nematic and smectic ordering. *J. Chem. Phys.* **1982**, *76*, 1708–1711. [CrossRef]
21. Fung, B.M.; Poon, C.-D.; Gangoda, M.; Enwall, E.L.; Diep, T.A.D.; Bui, C.V. Nematic and smectic ordering of 4-n-octyl-4′-cyanobiphenyl studied by carbon-13 NMR. *Mol. Cryst. Liq. Cryst.* **1986**, *141*, 267–277. [CrossRef]
22. McMillan, W.L. Simple molecular model for the smectic A phase of liquid crystal. *Phys. Rev. A* **1971**, *4*, 1238–1246. [CrossRef]
23. Ganzenmuller, G.C.; Patey, G.N. Charge ordering induces a smectic phase in oblate ionic liquid crystals. *Phys. Rev. Lett.* **2010**, *105*, 137801. [CrossRef]
24. Gorkunov, M.V.; Osipov, M.A.; Kapernaum, N.; Nonnenmacher, D.; Giesselmann, F. Molecular theory of smectic ordering in liquid crystals with nanoscale segregation of different molecular fragments. *Phys. Rev. E* **2011**, *84*, 051704. [CrossRef]
25. Klimavicius, V.; Gdaniec, Z.; Kausteklis, J.; Aleksa, V.; Aidas, K.; Balevicius, V. NMR and raman spectroscopy monitoring of proton/deuteron exchange in aqueous solutions of ionic liquids forming hydrogen bond: A role of anions, self-aggregation, and mesophase formation. *J. Phys. Chem. B* **2013**, *117*, 10211–10220. [CrossRef] [PubMed]
26. Every, H.A.; Bishop, A.G.; MacFarlane, D.R.; Oradd, G.; Forsyth, M. Transport properties in a family of dialkylimidazolium ionic liquids. *Phys. Chem. Chem. Phys.* **2004**, *6*, 1758–1765. [CrossRef]
27. Matveev, V.V.; Markelov, D.A.; Ievlev, A.V.; Brui, E.A.; Tyutyukin, K.V.; Lahderanta, E. Molecular mobility in several imidazolium-based ionic liquids according to data of H-1 and C-13 NMR relaxation. *Magn. Reson. Chem.* **2018**, *56*, 140–143. [CrossRef]
28. Frise, A.E.; Dvinskikh, S.V.; Ohno, H.; Kato, T.; Furo, I. Ion channels and anisotropic ion mobility in a liquid-crystalline columnar phase as observed by multinuclear NMR diffusometry. *J. Phys. Chem. B* **2010**, *114*, 15477–15482. [CrossRef] [PubMed]
29. Frise, A.E.; Ichikawa, T.; Yoshio, M.; Ohno, H.; Dvinskikh, S.V.; Kato, T.; Furo, I. Ion conductive behaviour in a confined nanostructure: NMR observation of self-diffusion in a liquid-crystalline bicontinuous cubic phase. *Chem. Commun.* **2010**, *46*, 728–730. [CrossRef]
30. Cifelli, M.; Domenici, V.; Kharkov, B.B.; Dvinskikh, S.V. Study of translational diffusion anisotropy of ionic smectogens by NMR diffusometry. *Mol. Cryst. Liq. Cryst.* **2015**, *614*, 30–38. [CrossRef]

31. Saielli, G. MD simulation of the mesomorphic behaviour of 1-hexadecyl-3-methylimidazolium nitrate: Assessment of the performance of a coarse-grained force field. *Soft Matter* **2012**, *8*, 10279–10287. [CrossRef]
32. Saielli, G.; Voth, G.A.; Wang, Y.T. Diffusion mechanisms in smectic ionic liquid crystals: Insights from coarse-grained MD simulations. *Soft Matter* **2013**, *9*, 5716–5725. [CrossRef]

crystals

MDPI

Article

Comparison of the Mesomorphic Behaviour of 1:1 and 1:2 Mixtures of Charged Gay-Berne GB(4.4,20.0,1,1) and Lennard-Jones Particles

Tommaso Margola [1], Katsuhiko Satoh [2],* and Giacomo Saielli [3],*

[1] Department of Chemical Sciences of the University of Padova, via Marzolo, 1, I-35131 Padova, Italy; tommaso.margola@gmail.com
[2] Department of Chemistry, Osaka Sangyo University, Daito, Osaka 574-8530, Japan
[3] CNR Institute on Membrane Technology, Unit of Padova, via Marzolo, 1, I-35131 Padova, Italy
* Correspondence: ksatoh@las.osaka-sandai.ac.jp (K.S.); giacomo.saielli@unipd.it (G.S.)

Received: 1 September 2018; Accepted: 19 September 2018; Published: 20 September 2018

Abstract: We present a Molecular Dynamics study of mixtures of charged Gay-Berne (GB) ellipsoids and spherical Lennard-Jones (LJ) particles as models of ionic liquids and ionic liquid crystals. The GB system is highly anisotropic (GB(4.4,20.0,1,1)) and we observe a rich mesomorphism, with ionic nematic and smectic phases in addition to the isotropic mixed phase and crystalline phases with honeycomb structure. The systems have been investigated by analyzing the orientational and translational order parameters, as well as radial distribution functions. We have directly compared 1:1 mixtures, where the GB and LJ particles have a charge equal in magnitude and opposite in sign, and 1:2 mixtures where the number of LJ particles is twice as large compared to the GB and their charge half in magnitude. The results highlight the role of the long-range isotropic electrostatic interaction compared to the short-range van der Waals anisotropic contribution, and the effect of the stoichiometry on the stability of ionic mesophases.

Keywords: ionic liquids; liquid crystals; ionic liquid crystals; molecular dynamics

---

1. Introduction

Ionic liquid crystals (ILC) are interesting materials composed of ions, as are their analogues ionic liquids (IL), and exhibiting thermotropic mesomorphism as non-ionic liquid crystals (LC). A comprehensive review on this subject has been published [1] where the synthesis and technological applications of ILCs have been thoroughly discussed. ILCs are normally composed by organic nitrogen cations, like imidazolium [2], pyridinium [3,4], bipyridinium (also known as viologens) [5–8], guanidinium [9–12], pyrrolidinium [13], to mention the most common ones, and inorganic anions like halides, tetrafluoroborate, hexafluorophosphate, bis(trifluoromethanesulfonyl)imide (also known as bistriflimide, or Tf_2N^-). A remarkable feature of ILCs is that almost all known compounds exhibit a smectic A, or even a more ordered-like smectic B and T, type of LC phase. Very few examples of an ionic nematic phase have been reported in the literature [14–16]. The reason of this behavior can be traced back to the structure of the cations: this is usually formed by a charged moiety and a relatively long alkyl chain and these two parts tend to micro-segregate, therefore promoting the formation of layered, that is smectic, phases instead of isotropic phases. In fact, such micro-segregation is well-known also to affect the structure of the isotropic phase of ILs [17]: Systems with relatively short alkyl chains do not form ILC phases, nonetheless a non-homogeneous structure of the isotropic phase of the IL has been observed first from MD simulations [18–20] and later confirmed experimentally [21]. A detailed discussion about the structure and properties of ILs can be found in the literature [22]. The relationship between the micro-segregation in the isotropic phase of ILs and the emergence of ionic LC phases has been the subject of both theoretical [23], computational [24,25] and experimental works [26–28]. In particular, Nelyubina et al. [29] found a correlation between the ratio of cation/anion volumes and the minimum chain length of 1-alky-3-methylimidazolium salts for which an ionic LC phase is observed: the larger the anion, the longer the alkyl chain needs to be in order for the compound to exhibit an ILC phase. However, one of the main issues related to ILC structure, that is the lack of a stable, and with a wide temperature range of existence, ionic nematic phase, is still an open question. Interestingly, in [1] the authors also stressed the lack of systematic computational and theoretical investigations of ILCs that would help to establish a set of structure-properties relationships in order to guide the synthesis of compounds potentially able to display an ionic nematic phase. Recent fully atomistic simulations have been focused on the popular methylimidazolium salts [30,31].

In this article we are continuing our investigation of the phase diagram of mixtures of charged ellipsoidal Gay-Berne (GB) particles and spherical Lennard-Jones (LJ) particles as models of ILCs. A preliminary account concerning 1:1 mixtures of particles with opposite charges, $+q^*$ (the asterisk indicates scaled quantities, see below) for the GB and $-q^*$ for the LJ, and using the parameterization of the GB particles proposed by Bates and Luckhurst [32], was reported recently [33]. A more extensive investigation of the phase diagram of 1:1 mixtures using the GB parameterization of Berardi et al. has been also published [34]. Here, for the first time, we will compare the results obtained for 1:1 mixtures and 1:2 mixtures (still using the parameterization of Bates and Luckhurst), that is systems where the GB particles bear a positive $+2q^*$ charge while the LJ particles, twice in number, bear a charge of $-q^*$. Hereafter, the systems will be identified simply by the value of q^* and the stoichiometry of the mixture. By keeping all other interaction parameters fixed, we will compare three different packing fractions ($\eta^* = 0.371$, $\eta^* = 0.428$ and $\eta^* = 0.514$) thus highlighting and singling out which is the effect of the stoichiometry on the stability of ionic liquid crystal phases.

2. Materials and Methods

Simulations were run with the LAMMPS software (version 28/09/2016) package [35]. We used cubic boxes with periodic boundary conditions in the NVT (constant number of particles, volume and temperature) ensemble with 5488 particles (2744 GB and 2744 LJ) for the 1:1 mixtures and 5184 particles (1728 GB and 3456 LJ) for the 1:2 mixtures. We tested the box size effect by running a set of simulations for the 1:1 system at a packing fraction of 0.514 (one of the packing fractions studied, see below) and no charge with a box of 10,832 particles and the results of the energy vs scaled temperature showed negligible differences with the system of 5488 particles. Moreover, box size effects were also extensively tested in [34] and we observed negligible effects for a number of particles larger than 4000. At least 500,000 time steps have been used for equilibration, followed by 500,000 time steps of production runs; longer equilibration were run close to the transition points.

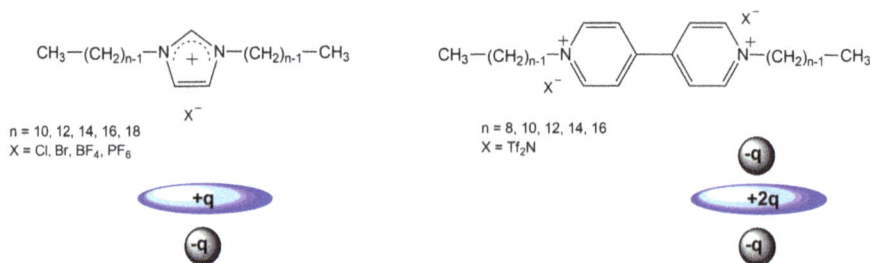

Figure 1. (top) Structural formula of some ionic liquid crystals (ILCs) with 1:1 stoichiometry (left, Reference [36,37]) and 1:2 stoichiometry (right, Reference [5,38–40]). (Bottom) schematic representation of the ion pairs of Gay–Berne (GB) and Lennard-Jones (LJ) particles used in the simulations.

The type of Gay-Berne potential selected for our simulations is that one proposed by Bates and Luckhurst [32]. This choice is dictated by the fact that such parameterization, compared to the parameterization of Berardi et al. [41], has a larger length to breadth ratio (4.4 instead of 3) and a larger side-by-side to end-to-end ratio of the interaction potential (20.0 instead of 5.0). For these reasons we expect it to be closer to very elongated molecules, such as the imidazolium and especially the viologens in Figure 1, with alkyl chains of several carbon atoms. It is worth mentioning that the relatively long and flexible alkyl chains of the real systems are not included in the model potential; however, this limitation is the same as encountered when the pure GB potential is used as a model to describe non-ionic liquid crystals, for example cyanobiphenyls, that very often have an alkyl chain of several carbon atoms. Nonetheless, the lack of alkyl chains in the GB potential has not limited its application to the study of LCs. The GB potential is shown in Equation (1): it is essentially a Lennard-Jones potential with a constant distance and well depth between the particles that depends on the relative orientations of the two ellipsoids, u_i and u_j, and the orientation of the vector connecting their center of mass, r_{ij} [42].

$$U_{GB} = U(u_i, u_j, r_{ij}) = 4\varepsilon_0 \varepsilon(u_i, u_j, r_{ij}) \left[\left(\frac{\sigma_0}{r_{ij} - \sigma(u_i, u_j, r_{ij}) + \sigma_0} \right)^{12} - \left(\frac{\sigma_0}{r_{ij} - \sigma(u_i, u_j, r_{ij}) + \sigma_0} \right)^{6} \right], \tag{1}$$

The parameters σ_0 and ε_0 are used to scale the distance and the energy, respectively, for all interactions, which are the GB-GB, the LJ-LJ and the mixed GB-LJ interactions. The LJ potential used for the sphere interaction, in fact, can be seen as the spherical limit of the GB potential in Equation (1) that can be obtained by setting $\varepsilon(u_i, u_j, r_{ij}) = 1$ and $\sigma_0(u_i, u_j, r_{ij}) = \sigma_0$. The mixed potential is based on the interaction potential between two biaxial particles as implemented in the software package LAMMPS [35] based on the theoretical derivation of [43,44]. Since we are interested in understanding the effect of the stoichiometry on the stability of ionic mesophases we keep all other interaction parameters fixed (except the charge). Throughout the manuscript we will make use of scaled quantities, defined with respect to the distance and interaction scaling parameters of the GB potential as follows:

$$
\begin{aligned}
&\text{the scaled potential energy, } U^* = \frac{U}{\varepsilon_0}; \\
&\text{the scaled distance, } r^* = \frac{r}{\sigma_0}; \\
&\text{the scaled volume, } V^* = \frac{V}{\sigma_0^3}; \\
&\text{the scaled number density, } \rho^* = \frac{N}{V^*}; \\
&\text{the scaled packing fraction, } \eta^* = \frac{N V_m^*}{V^*}; \text{ where } V_m^* \text{ is the scaled molecular volume} \\
&\text{the scaled temperature, } T^* = \frac{k_B T}{\varepsilon_0}; \\
&\text{the scaled time, } t^* = t \sqrt{\frac{\varepsilon_0}{m \sigma_0}}; \\
&\text{the scaled charge, } q^* = \frac{q}{\sqrt{\varepsilon_0 \sigma_0}},
\end{aligned}
\tag{2}
$$

The two initial boxes were prepared as follows: the initial 1:1 box was obtained by first creating a unit cell of size $r^* = 5$ containing an ion pair (one GB and one LJ particles) manually placed inside at random position and orientation, avoiding overlap; then the unit cell was replicated 14 times along x, y and z direction of the box, thus $14 \times 14 \times 14 = 2744$ ion pairs = 5488 particles. For the initial 1:2 box the same unit cell was filled with one GB and two LJ particles and then replicated 12 times along x, y and z, thus $12 \times 12 \times 12 = 1728$ ion triplets = 5184 particles. The boxes were first equilibrated for 1000 t-steps at this very low density and $T^* = 8.0$ to completely randomize positions and orientations, then the box size and particle positions were rescaled at the desired packing fraction at the same relatively high temperature of $T^* = 8.0$. The obtained configurations were used as starting points for the cooling runs where the box at each temperature was equilibrated starting from the box obtained at the previous run at higher temperature. Cooling runs were then followed by heating runs to check for possible hysteresis.

Electrostatic interactions were handled using the particle-particle particle-mesh solver [45] setting a precision of 10^{-6} on the calculated forces. The cut-off for the short-range interaction was set to 10 scaled units and the temperature was controlled by a Nose-Hoover thermostat [46].

The results of the MD simulations will be analyzed by calculating the orientational order parameter of the GB particles, $\langle P_2 \rangle = \langle \frac{1}{2}(3\cos^2\theta - 1) \rangle$, where the brackets indicate an ensemble average over the GB particles only. θ is the angle between the director of the phase and the long axis of the GB particle. In addition to the orientational order, we also calculate the translational order parameter as $\tau = |exp(2i\pi r^*/d^*)|$, where d^* is the scaled layer thickness of the smectic phase. The calculation of follows the procedure of [32]. In the evaluation of $\langle \tau \rangle$ we considered an average value among the 10 last configurations of the cooling run. In the snapshots, obtained with the software package QMGA [47], GB particles are colour-coded based on the orientation with respect to the director.

3. Results

We will present a direct comparison of the phase behavior of the two mixtures, with 1:1 and 1:2 stoichiometry. In order to correctly compare the systems, we set the scaled packing fraction, η^*, to the same value, thus the number densities differ, as indicated in Table 1.

Table 1. Systems investigated in this work.

	ρ^*(1:1)	ρ^*(1:2)	η^*
Case #1	0.261	0.332	0.371
Case #2	0.303 [a]	0.383	0.428
Case #3	0.363 [a]	0.460	0.514

[a] System investigated in [33].

3.1. Comparison of 1:1 and 1:2 Stoichiometry for a Packing Fraction 0.371

In Figure 2 we report the dependence of the orientational order parameter $\langle P_2 \rangle$ and the translational order parameter $\langle \tau \rangle$ on the scaled temperature T^* for systems with different scaled charges q^*. We recall here that a given value of q^* means charges of $+q^*$ and $-q^*$ for GB and LJ, respectively, for the 1:1 mixture; while it means $+2q^*$ and $-q^*$ for GB and LJ, respectively, for the 1:2 mixture. Thus, for clarity, we will use, hereafter, the explicit indication of both charges for the GB:LJ mixtures, as $q_{GB}^*/-q_{LJ}^*$. So, for example, a value of $q^* = 1.00/-0.50$ would indicate a 1:2 GB:LJ mixture with $q_{GB}^* = +1.00$ and $q_{LJ}^* = -0.50$; while a notation of $q^* = 0.50/-0.50$ would represent a 1:1 mixture with $q_{GB}^* = +0.50$ and $q_{LJ}^* = -0.50$.

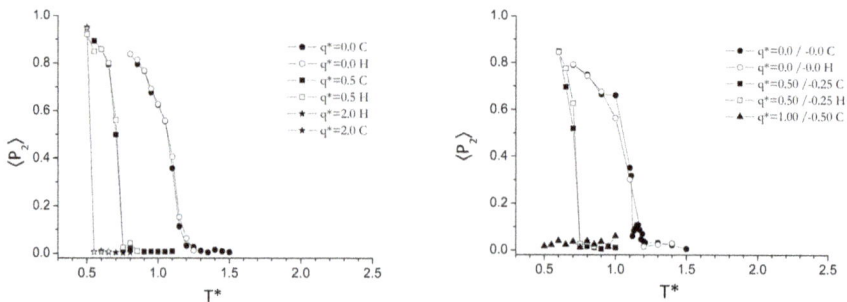

Figure 2. *Cont.*

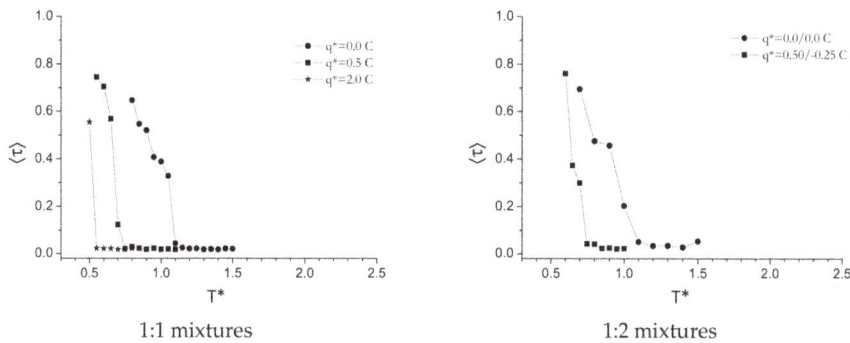

1:1 mixtures   1:2 mixtures

Figure 2. Dependence of the orientational order parameter, $\langle P_2 \rangle$, (top panels) and translational order parameter, $\langle \tau \rangle$, (bottom panels) on the scaled temperature, T^*. The left panels refer to the 1:1 stoichiometry; the right panels to the 1:2 stoichiometry. Packing fraction $\eta^* = 0.371$. H and C refer to heating and cooling runs, respectively.

For the non-charged systems, $q^* = 0.0/0.0$, we observe a similar behavior for both stoichiometries: there is a discontinuity in the orientational order parameter are at $T^* = 1.15$ and $T^* = 1.10$ for the 1:1 and 1:2 mixtures, respectively, while the translational order parameter remains zero until the temperature is lowered down to $T^* = 1.05$ and $T^* = 1.00$, for the 1:1 and 1:2 mixtures, respectively. Therefore, we can conclude that a narrow range of ionic nematic phase is present, with a temperature range of about 0.1 scaled unit, and a thermal range slightly at lower temperature for the 1:2 stoichiometry. Then, below $T^* = 1.05$ and $T^* = 1.00$, a smectic phase is formed.

However, as we can expect for a mixture of non-charged particles of very different shape, a phase separation occurs. This phenomenon has been studied by several authors, e.g., for mixtures of hard rod-core particles [48] and soft and non-charged GB/LJ mixtures [49]. The observed phase separation can be traced back to entropy effects related with the excluded volume and is generally stronger the more the particle's shapes and volumes differ.

Such phase separation can be clearly seen in the snapshots reported in Figure 3. Interestingly, for the 1:2 mixture, the phase separation occurs before the transition of the GB particles into a LC phase, see Figure 3c, left snapshot.

A more interesting phase behavior is observed once the charge is switched on. First, we observe a general decrease of the phase-transition temperatures, and the larger the charge the lower is the temperature where the isotropic mixed state undergoes a transition to a more ordered phase. For a charge $q^* = 0.50/-0.50$ (1:1) and $q^* = 0.50/-0.25$ (1:2) the transition from isotropic to a LC phase with an incipient smectic ordering (since $\langle \tau \rangle = 0.1$) is at about $T^* = 0.70$, in both cases (1:1 and 1:2 mixtures), with an overall increase of the stability of the isotropic mixed phase of about $\Delta T^* = 0.40$, with respect of the non-charge case. Inspection of the values of $\langle P_2 \rangle$ and $\langle \tau \rangle$ reveals that at the first point, on cooling the temperature, where $\langle P_2 \rangle \neq 0$ we also have $\langle \tau \rangle \neq 0$, meaning that the phase formed is of smectic type and a nematic phase may exist only in a temperature range smaller than 0.05, the step in the temperature used during the heating and cooling runs. Snapshots of the simulations boxes can be seen in Figure 3.

For a charge $q^* = 1.0/-1.0$ (1:1) we observe again a single transition from the isotropic mixed phase to the smectic phase for the 1:1 stoichiometry. The transition temperature is now much lower, at $T^* = 0.50$; in contrast the 1:2 stoichiometric mixture does not exhibit any mesophase for $q^* = 2.0/-1.0$ (1:2) and the system remains isotropic and mixed until the lowest temperature investigated, that is $T^* = 0.50$, where a transition to a highly ordered smectic phase is observed. Therefore, increasing the charge leads to a destabilization of the ionic LC phases, as already observed for a different GB/LJ system with a 1:1 stoichiometry [34].

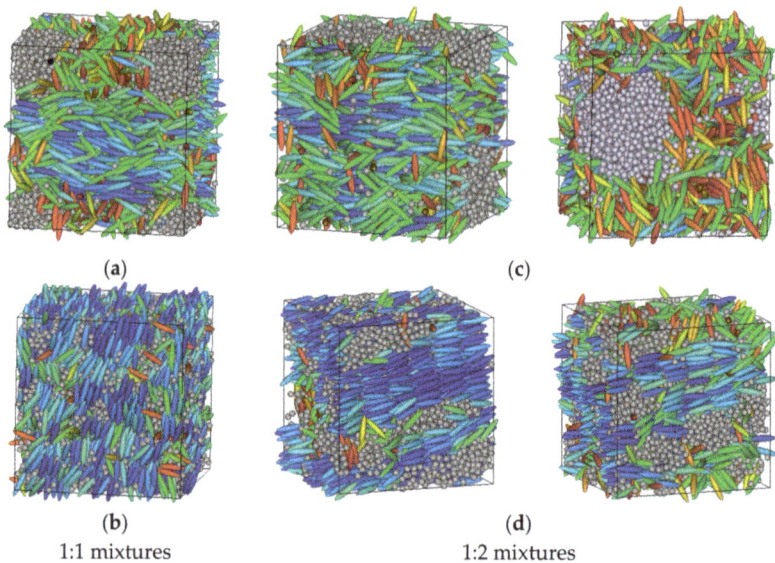

Figure 3. (**a**) 1:1 mixture: phase-separated nematic GB and isotropic LJ: $q^* = 0.0/-0.0$, $T^* = 1.10$; (**b**) 1:1 mixture: ionic LC phase with incipient smectic layering of the GB particles: $q^* = 0.50/-0.50$, $T^* = 0.65$. (**c**) 1:2 mixture: phase separated GB and LJ particles: $q^* = 0.0/-0.0$, $T^* = 1.10$ (left snapshot) and $T^* = 1.30$ (right snapshot); (**d**) 1:2 mixture: microphase segregation GB and LJ particles: $q^* = 0.50/-0.25$, $T^* = 0.65$ (left snapshot) and $T^* = 0.70$ (right snapshot), with GB particles arranged in layers. Packing fraction $\eta^* = 0.371$.

3.2. Comparison of 1:1 and 1:2 Stoichiometry for a Packing Fraction 0.428

In Figure 4 we show the orientational and translational order parameters for the two mixtures with 1:1 and 1:2 GB:LJ stoichiometry at a packing fraction $\eta^* = 0.428$.

As we can see, increasing the density of the systems destabilize the isotropic mixed phase. For the non-charged system, the first transition on cooling down the temperature occurs at $T^* = 1.80$ and $T^* = 2.0$, for the 1:1 and 1:2 mixtures, respectively. Such a transition is accompanied by a phase separation with a phase composed by GB particles in the LC phase and isotropic LJ particles. The thermal range of stability of such bi-phasic system is rather large since the translational order parameter becomes different from zero only at T^* around 1.4 for both systems.

As already observed for the previous less-dense system, the presence of opposite charges again increases the stability of the isotropic mixed phase by opposing the phase separation. For a charge $q^* = 0.50/-0.50$ (1:1) and $q^* = 0.50/-0.25$ (1:2) the transition of the system from an isotropic into a nematic phase is about $T^* = 1.3$ and $T^* = 1.1$ for the 1:1 and 1:2 mixtures, respectively. Moreover, while the 1:1 mixture has a non-negligible thermal range of existence of the nematic phase (ca. $\Delta T^* \approx 0.4$), the 1:2 systems almost immediately goes into a smectic phase. For larger charges, we did not observe a clear ILC phase, rather a direct transition into a crystal lattice of GB particles as shown in Figure 5d. As we can see the GB particles arrange in triangular channel, or nanotubes whose walls are made by a side-by-side packing of the ellipsoids. The inside of the nanotubes is filled with an ordered distribution of LJ particles. The nano-channels are arranged into a honeycomb structure.

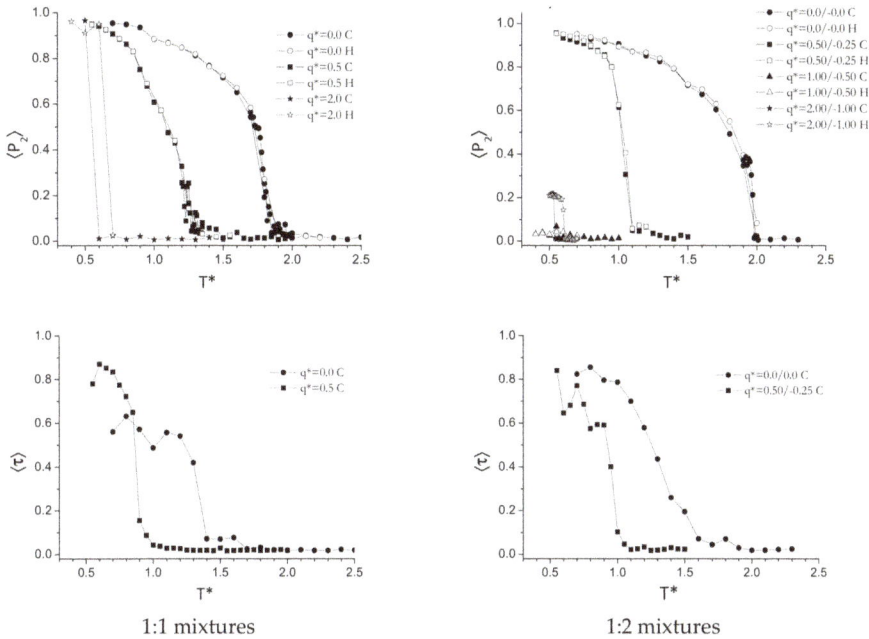

Figure 4. Dependence of the orientational order parameter, $\langle P_2 \rangle$, (top panels) and translational order parameter, $\langle \tau \rangle$, (bottom panels) on the scaled temperature, T^*. The left panels refer to the 1:1 stoichiometry; the right panels to the 1:2 stoichiometry. Packing fraction $\eta^* = 0.428$. H and C refer to heating and cooling runs, respectively

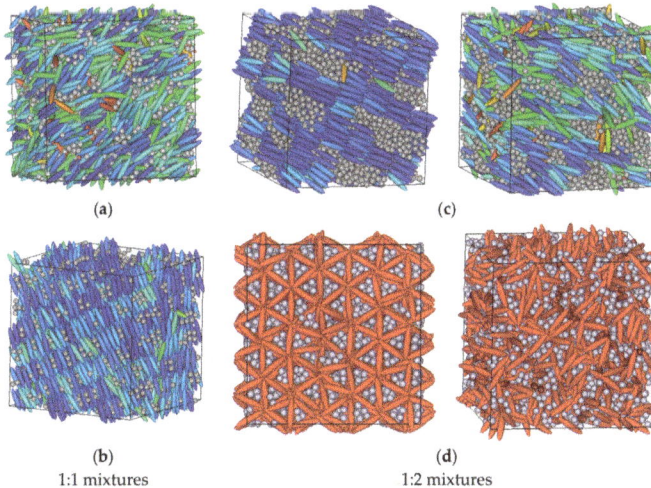

Figure 5. (**a**) 1:1 mixture: ionic nematic phase: $q^* = 0.50/-0.50$, $T^* = 0.90$; (**b**) 1:1 mixture: highly ordered crystal/smectic phase: $q^* = 2.00/-2.00$, $T^* = 0.50$. (**c**) 1:2 mixture: microphase segregated GB and LJ particles: $q^* = 0.50/-0.25$, $T^* = 0.55$ (left snapshot) and $T^* = 1.00$ (right snapshot); (**d**) 1:2 mixture: microphase segregation GB and LJ particles: $q^* = 2.00/-1.00$, $T^* = 0.50$ (left snapshot) and $T^* = 0.55$ (right snapshot), with GB particles arranged into an hexagonal structure hosting LJ particles inside the channels. We do not use colour-coding here since there is not a director of the phase. Packing fraction $\eta^* = 0.428$.

3.3. Comparison of 1:1 and 1:2 Stoichiometry for a Packing Fraction 0.514

In Figure 6 we show the orientational and translational order parameters for the mixtures at the packing fraction of 0.514.

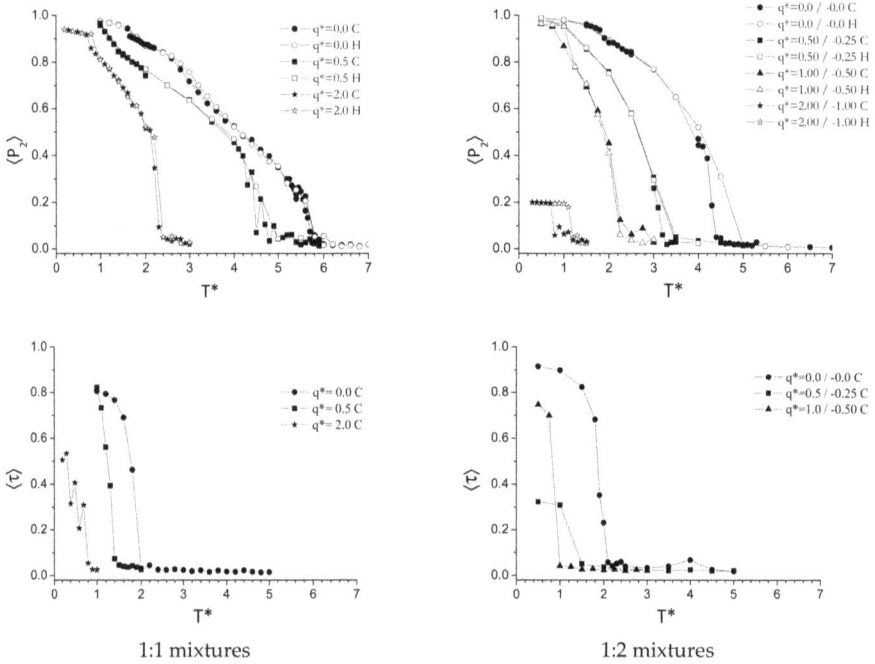

1:1 mixtures   1:2 mixtures

Figure 6. Dependence of the orientational order parameter, $\langle P_2 \rangle$, (top panels) and translational order parameter, $\langle \tau \rangle$, (bottom panels) on the scaled temperature, T^*. The left panels refer to the 1:1 stoichiometry; the right panels to the 1:2 stoichiometry. Packing fraction $\eta^* = 0.514$. H and C refer to heating and cooling runs, respectively.

As we can see, following the trend already observed, increasing the packing fraction leads to a stabilization of the ordered phases and a destabilization of the isotropic mixture. In particular, for the non-charged 1:1 system a nematic phase of the GB particles is found between about $T^* = 2.0$ and $T^* = 5.9$, although as for the previous cases, the transition is concomitant with a phase separation. In contrast, for a scaled charge $q^* = 0.50/-0.50$ we observe a homogeneous ionic nematic phase in the range between $T^* = 1.4$ and $T^* = 4.8$. A relatively wide range of ionic nematic phase is also observed for the higher charge of $q^* = 2.00/-2.00$, extending from $T^* = 0.9$ to $T^* = 2.4$.

Comparing these results with those obtained from the 1:2 mixture we note a significant destabilization of the LC phase in favour of a mixed isotropic phase. For the non-charged 1:2 system the first transition, on cooling down the temperature, occurs around $T^* = 4.5$ (we note some hysteresis), that is 1.4 units of scaled temperature lower than the corresponding 1:1 mixture. Also, the charged systems exhibit much lower transition temperatures: for example, both the 1:2 mixtures with charge $q^* = 0.50/-0.25$ and $q^* = 1.00/-0.50$ have a transition from the isotropic into the ionic nematic phase at $T^* = 3.2$ and $T^* = 2.2$, respectively, much below the value of $T^* = 4.8$ of the 1:1 mixture with charge $0.50/-0.50$. Finally, as noted already for the lower packing fraction of 0.428, the 1:2 mixture, when the charge is above a certain value does not exhibit any LC phase, rather a crystalline structure where the GB particles form channels of a hexagonal structure hosting the spherical LJ particles.

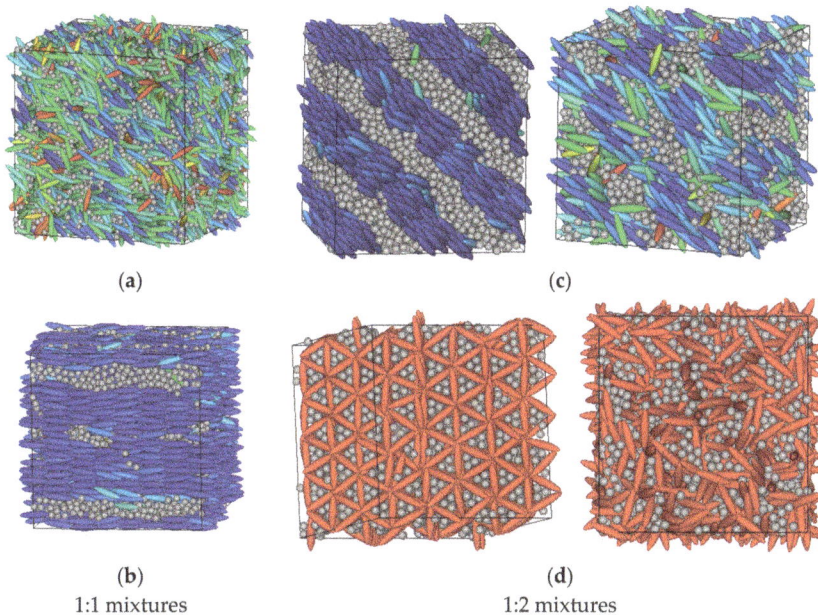

(a)   (c)

(b)   (d)

1:1 mixtures   1:2 mixtures

Figure 7. (**a**) 1:1 mixture: ionic nematic phase: $q^* = 0.50/-0.50$, $T^* = 4.00$; (**b**) 1:1 mixture: microphase segregated ionic smectic phase: $q^* = 0.50$, $T^* = 1.00$. (**c**) 1:2 mixture: ionic nematic phase: $q^* = 0.50/-0.25$, $T^* = 1.00$ (left snapshot) and $T^* = 2.00$ (right snapshot); (**d**) 1:2 mixture: microphase segregation GB and LJ particles: $q^* = 2.00/-1.00$, $T^* = 0.30$ (left snapshot) and $T^* = 0.70$ (right snapshot), with GB particles arranged to create a hexagonal structure hosting LJ particles inside the channels. Packing fraction $\eta^* = 0.514$.

We mention here that very similar aggregation patterns as found in Figure 7c have been recently observed experimentally in mixtures of oppositely charged colloidal particles [50,51]. In that case, the formation of alternating layers of positive and negative particles was obtained by imposing an oscillating electric field. In Figure 7d, the GB particles also arrange in triangular channels, or nanotubes, by a side-by-side packing of the ellipsoids. The inside of the nanotubes is filled with a few LJ particles. The nano-channels are arranged into a honeycomb structure similarly to the system with $\eta^* = 0.428$.

4. Discussion

In Table 2 we report a summary of the phases and transition temperatures obtained by the combined observation of the orientational and translational order parameters, $\langle P_2 \rangle$ and $\langle \tau \rangle$, respectively, recalling that they are calculated considering only the GB particles' orientation and positions. The assignment is made as follows: if both $\langle P_2 \rangle = 0$ and $\langle \tau \rangle = 0$ the phase is isotropic; if $\langle P_2 \rangle \neq 0$ and $\langle \tau \rangle = 0$ the phase is nematic; finally, if both $\langle P_2 \rangle \neq 0$ and $\langle \tau \rangle \neq 0$ the phase is smectic. Inspection of the radial distribution functions (see supporting information) and snapshots is used to confirm the assignment as well as to assign crystal phases. Therefore, only structural properties of the GB particles have been considered for the phase assignment without explicitly considering the amount of clustering and micro-segregation or mixing/demixing of the GB/LJ particles and also without considering dynamic properties and possibly the formation of glassy states. A more detailed description of the phases can be obtained by inspection of the snapshots in Figures 3, 5 and 7. Moreover, in Figure 8 we show the thermal range of stability of the phases of the various systems as histograms.

Table 2. Systems investigated in this work.

	GB:LJ = 1:1		GB:LJ = 1:2		
$|q^*|$	Phases	q^*	Phases	η^*
0.0 [a]	Sm–1.05–N–1.15–Iso	0.0/0.0 [a]	Sm–1.10–Iso	
0.5	Sm–0.70–N–0.75–Iso	0.50/−0.25	Sm–0.70–Iso	0.371
2.0	Cr–0.50–Iso	1.00/−0.50	Iso	
0.0 [a]	Sm–1.35–N–1.84–Iso	0.0/0.0 [a]	Sm–1.5–N–2.0–Iso	
0.5	Sm–0.90–N–1.25–Iso	0.50/−0.25	Sm–1.0–N–1.10–Iso	0.428
2.0	Cr–0.65–Iso	1.00/−0.50	Cr–0.65–Iso	
		2.00/−1.00	Cr–0.55–Iso	
0.0 [a]	Sm–1.9–N–5.7–Iso	0.0/0.0 [a]	Sm–2.05–N–4.30–Iso	
0.5	Sm–1.4–N–4.9–Iso	0.50/−0.25	Sm–1.0–N–3.1–Iso	
2.0	Sm–0.75–N–2.25–Iso	1.00/−0.50	Sm–0.75–N–2.2–Iso	0.514
		2.00/−1.00	Cr–1.0–Iso	

[a] For non-charged systems we always observe a phase separation between GB and LJ particles.

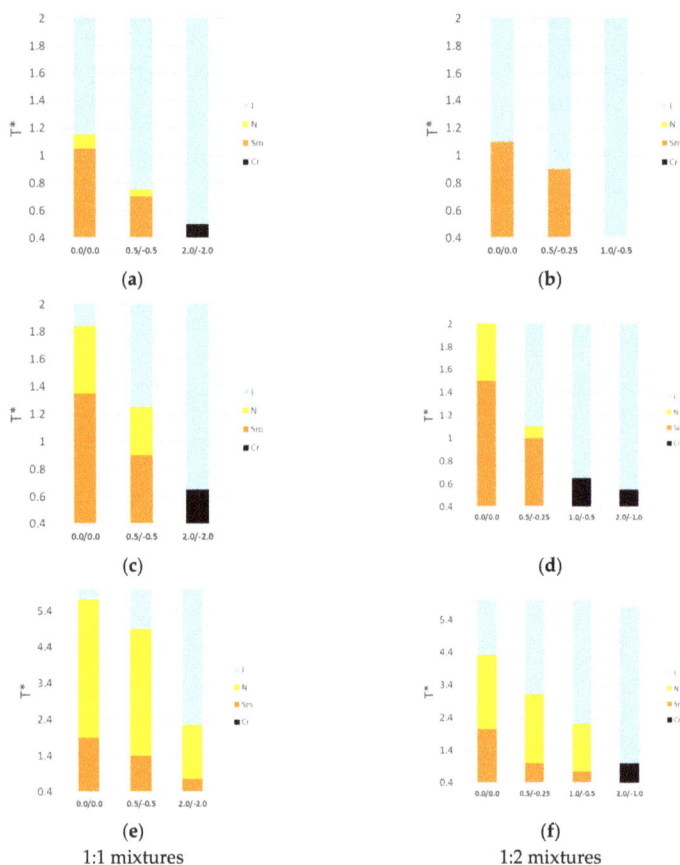

Figure 8. Temperature range of stability of the various systems investigated. (**a**): packing fraction $\eta^* = 0.371$, GB:LJ = 1:1; (**b**) packing fraction $\eta^* = 0.371$, GB:LJ = 1:2; (**c**): packing fraction $\eta^* = 0.428$, GB:LJ = 1:1; (**d**) packing fraction $\eta^* = 0.428$, GB:LJ = 1:2; (**e**): packing fraction $\eta^* = 0.514$, GB:LJ = 1:1; (**f**) packing fraction $\eta^* = 0.514$, GB:LJ = 1:2. Note the different T^* scale for the highest packing fraction (**e**,**f**).

The main differences between the 1:1 and 1:2 stoichiometries can be summarized as follows: by increasing the mole fraction of spherical LJ particles (and changing the relative charge accordingly, in order to keep the system electrically neutral) we note a significant destabilization of the LC phases; for the lowest packing fraction investigated we note that the nematic range appears smaller in the non-charged 1:2 mixtures while even the smectic phase is suppressed when the charge is increased to $q^* = 2.00/-1.00$. Increasing the density confirms this trend and, while the charged 1:1 mixtures do exhibit a sequence of ionic nematic and ionic smectic phases, for the analogous 1:2 mixtures these are destabilized, and interesting hexagonal arrangements appear made of the highly anisotropic, and positively charged, GB particles hosting the negatively charged LJ spheres. This arrangement maximizes the side-by-side van der Waals interaction of the GB particles and, at the same time, the electrostatic energetic cost due to the repulsion between cations is compensated by the attractive interaction with the anions inside the channels. The GB interaction is, in fact, highly anisotropic, being 20 times larger, in the side-by-side configuration, than both the end-to-end interaction and the LJ-LJ interaction, so it plays a major role in the phase structure of the systems, favouring a smectic type of phase for the GB particles.

The honeycomb arrangement observed at relatively high packing fractions, high charge and low temperatures can be understood as a fortuitous matching of the relative length of the GB particles and size of the LJ particles. In Figure 9 we show a very schematic and bi-dimensional representation of the triangular unit composing the honeycomb structure. This is an oversimplified geometrical model just to highlight the importance of the right length of the GB particles; thus, we ignore, for the sake of simplicity, the overlap at the ends of each ellipsoid. The side of the equilateral triangle, in such an arrangement, is $l = 4.4$, while the radius of the LJ sphere is $r_{LJ} = 0.5$. A simple calculation shows that the available area inside, that is the area of the triangle excluding the internal area occupied by the ellipsoids, is $l \cdot \cos(30°) \cdot l/2 - l \cdot r_{LJ} \cdot \pi \cdot 1.5 = 3.1995$. The last factor 1.5, accounts for the fact that each triangle contains only half of each ellipsoid, that is 1.5 in total (besides, it contains also 3 spheres, thus keeping the correct stoichiometry of 1:2). The area of the three spheres is only 2.3561945, that is less than the available area inside the triangle. Therefore, three LJ spheres can be accommodated inside the honeycomb structure for this particular GB.

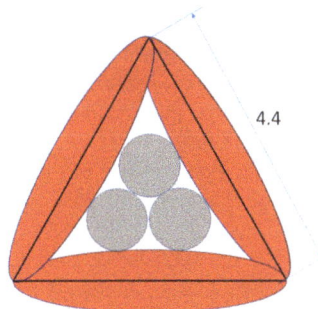

Figure 9. Schematic structural unit of the honeycomb arrangement observed for some systems at high packing fraction, high charge and low temperature.

We might expect that for a more anisotropic GB particle a different structure might be formed, e.g., of square shape. This behavior is very close to what was extensively studied by Tschierske and co-workers in several papers, see for example [52,53]. The systems investigated by the authors consists of the so-called T-shape mesogens, generally non ionic, where one or more flexible tails are linked to the middle of a rigid core. Various types of honeycomb structures with different symmetries can be observed depending on the ratio between the length of the rigid core and the lateral chains. In our mixtures made by ellipsoids and spheres, the size and shape anisotropy of the ellipsoids and the size of the spheres are the most important factors leading to the formation of the interesting geometrical structures such as the honeycomb and square shapes. The second important factor is weak ionicity, because the strong charge would cause strong ion pairing by the GB and LJ oppositely charged particles, thus avoiding the macroscopic phase separation as described above.

A second interesting effect is the stronger tendency to phase-separate of the 1:2 non-charged mixtures, compared to the analogous non-charged 1:1 mixtures. Because of the strong shape anisotropy of the GB particles (breadth-to-length ratio of 4.4), excluded volume effects are quite significant and the mixing is observed only at relatively high temperatures, or low densities. For the 1:2 mixture we observe phase separation, on cooling the temperature, even before the GB particles have a transition into a LC phase.

Therefore, ionic nematic phases can be formed in mixtures of particles of different shapes, oppositely charged; however, a fine tuning of the van der Waals (vdW) vs electrostatic (ES) contributions is necessary. In particular, doubling the charge of the ellipsoidal particles compared to that of the spheres (thus doubling the number of LJ particles to keep the system neutral) is detrimental to the stability of ionic LC phases. The fine tuning of vdW and ES interaction would be realized by designed molecular structure by steric, flexibility, effective charge strength and so on. However, the studied mixture of GB and LJ particles proved the possibility of forming a nematic ionic liquid crystalline phase.

Finally, we note that a general effect of increasing the charge, for a given packing fraction, is to increase the stability of the isotropic mixed phase, and therefore to destabilize the ionic liquid mesophases. This result has been also experimentally confirmed recently by Laschat and co-workers [54]: the authors investigated a series of amino acid/crown ether ionic LCs and found that increasing the charge is detrimental for the stability of ionic LC phases.

5. Conclusions

To summarize, in this manuscript we have presented a direct comparison of 1:1 and 1:2 mixtures of charged GB and LJ particles as models of ionic liquids and ionic liquid crystals. Despite the simplicity of the model, we observe a rich polymorphism with isotropic, nematic, smectic and crystal phases of ionic type. Generally speaking, both increasing the charge above a certain limit, as well as increasing the ratio of LJ over GB particles, leads to a destabilization of the ionic mesophases. While tuning the charge in the simulations is very easy, this is not true for a real ionic compound. One possible approach could be to favour charge transfer from the anion to the cation. It is, in fact, well-known that charge transfer occurs in ionic liquids and the effective charge of the ions is around 0.7–$0.8e$ for some cation–anion combinations [55]. Another possibility is to increase the relative weight of π-π interactions and their anisotropy in the rigid core in order to make the electrostatic contribution less important. Finally, these results suggest that, in order to increase the thermal range of stability of ionic mesophases, especially if the elusive ionic nematic phase is sought, one possible strategy is to vary the stoichiometry in favour of the anisotropic GB cations, e.g., by synthesizing salts of 2:1 stoichiometry, with the cations bearing half the charge of the anions, and to increase the van der Waals forces compared to the electrostatic interaction by designing systems with large polarizability.

Supplementary Materials: The following are available online at www.mdpi.com/link, Figures S1–S20: radial distribution functions, g(r), of the systems investigated.

Author Contributions: G.S. and K.S. conceived and designed the simulations; T.M. performed the simulations; T.M., K.S. and G.S. analyzed the data; T.M., K.S. and G.S wrote the paper.

Funding: This research received no external funding.

Acknowledgments: Computational work has been carried out on the C3P (Computational Chemistry Community in Padua) HPC facility of the Department of Chemical Sciences of the University of Padua and on the HPC facilities of the CINECA consortium (Bologna, Italy) through the ISCRA projects HP10CCR1RQ and HP10CNVBJY.

Conflicts of Interest: The authors declare no conflict of interest.

References

1. Goossens, K.; Lava, K.; Bielawski, C.W.; Binnemans, K. Ionic liquid crystals: Versatile materials. *Chem. Rev.* **2016**, *116*, 4643–4807. [CrossRef] [PubMed]
2. Bowlas, C.J.; Bruce, D.W.; Seddon, K.R. Liquid-crystalline ionic liquids. *Chem. Commun.* **1996**, *14*, 1625–1626. [CrossRef]
3. Nusselder, J.J.H.; Engberts, J.B.F.N.; VanDoren, H.A. Liquid-crystalline and thermochromic behavior of 4-substituted 1-methylpyridinium iodide surfactants. *Liq. Cryst.* **1993**, *13*, 213–225. [CrossRef]
4. Ujiie, S.; Mori, A. Cubic mesophase formed by thermotropic liquid crystalline ionic systems-effects of polymeric counterion. *Mol. Cryst. Liq. Cryst.* **2005**, *437*, 1269–1275. [CrossRef]
5. Causin, V.; Saielli, G. Effect of asymmetric substitution on the mesomorphic behaviour of low-melting viologen salts of bis(trifluoromethanesulfonyl)amide. *J. Mater. Chem.* **2009**, *19*, 9153. [CrossRef]
6. Casella, G.; Causin, V.; Rastrelli, F.; Saielli, G. Viologen-based ionic liquid crystals: Induction of a smectic A phase by dimerisation. *Phys. Chem. Chem. Phys.* **2014**, *16*, 5048–5051. [CrossRef] [PubMed]
7. Casella, G.; Causin, V.; Rastrelli, F.; Saielli, G. Ionic liquid crystals based on viologen dimers: Tuning the mesomorphism by varying the conformational freedom of the ionic layer. *Liq. Cryst.* **2016**, *43*, 1161–1173. [CrossRef]
8. Asaftei, S.; Ciobanu, M.; Lepadatu, A.M.; Enfeng, S.; Beginn, U. Thermotropic ionic liquid crystals by molecular-assembly and ion pairing of 4,4'-bipyridinium derivatives and tris(dodecyloxy)benzenesulfonats in a non-polar solvent. *J. Mater. Chem.* **2012**, *22*, 14426–14437. [CrossRef]
9. Butschies, M.; Sauer, S.; Kessler, E.; Siehl, H.-U.; Claasen, B.; Fischer, P.; Frey, W.; Laschat, S. Influence of N-Alkyl Substituents and Counterions on the Structural and Mesomorphic Properties of Guanidinium Salts: Experiment and Quantum Chemical Calculations. *Chemphyschem* **2010**, *11*, 3752–3765. [CrossRef] [PubMed]
10. Sauer, S.; Saliba, S.; Tussetschlager, S.; Baro, A.; Frey, W.; Giesselmann, F.; Laschat, S.; Kantlehner, W. p-Alkoxybiphenyls with guanidinium head groups displaying smectic mesophases. *Liq. Cryst.* **2009**, *36*, 275–299. [CrossRef]
11. Sauer, S.; Steinke, N.; Baro, A.; Laschat, S.; Giesselmann, F.; Kantlehner, W. Guanidinium chlorides with triphenylene moieties displaying columnar mesophases. *Chem. Mater.* **2008**, *20*, 1909–1915. [CrossRef]
12. Butschies, M.; Frey, W.; Laschat, S. Designer ionic liquid crystals based on congruently shaped guanidinium sulfonates. *Chem. Eur. J.* **2012**, *18*, 3014–3022. [CrossRef] [PubMed]
13. Tao, J.; Zhong, J.; Liu, P.; Daniels, S.; Zeng, Z. Pyridinium-based ionic liquid crystals with terminal fluorinated pyrrolidine. *J. Fluor. Chem.* **2012**, *144*, 73–78. [CrossRef]
14. Goossens, K.; Nockemann, P.; Driesen, K.; Goderis, B.; Goerller-Walrand, C.; Hecke, K.; Van Meervelt, L.; Van Pouzet, E.; Binnemans, K.; Cardinaels, T. Imidazolium ionic liquid crystals with pendant mesogenic groups. *Chem. Mater.* **2008**, *20*, 157–168. [CrossRef]
15. Li, W.; Zhang, J.; Li, B.; Zhang, M.; Wu, L. Branched quaternary ammonium amphiphiles: Nematic ionic liquid crystals near room temperature. *Chem. Commun.* **2009**, *35*, 5269–5271. [CrossRef] [PubMed]
16. Ringstrand, B.; Jankowiak, A.; Johnson, L.E.; Kaszynski, P.; Pociecha, D.; Gorecka, E. Anion-driven mesogenicity: A comparative study of ionic liquid crystals based on the [closo-1-CB9H10](-) and [closo-1-CB11H12](-) clusters. *J. Mater. Chem.* **2012**, *22*, 4874–4880. [CrossRef]
17. Shi, R.; Wang, Y. Dual ionic and organic nature of ionic liquids. *Sci. Rep.* **2016**, *6*, 19612–19644. [CrossRef] [PubMed]
18. Urahata, S.M.; Ribeiro, M.C.C. Structure of ionic liquids of 1-alkyl-3-methylimidazolium cations: A systematic computer simulation study. *J. Chem. Phys.* **2004**, *120*, 1855–1863. [CrossRef] [PubMed]
19. Wang, Y.; Voth, G.A. Unique Spatial Heterogeneity in Ionic Liquids. *J. Am. Chem. Soc.* **2005**, *127*, 12192–12193. [CrossRef] [PubMed]
20. Canongia Lopes, J.N.A.; Pàdua, A.A.H. Nanostructural Organization in Ionic Liquids. *J. Phys. Chem. B* **2006**, *110*, 3330–3335. [CrossRef] [PubMed]
21. Triolo, A.; Russina, O.; Bleif, H.-J.; Di Cola, E. Nanoscale Segregation in Room Temperature Ionic Liquids. *J. Phys. Chem. B* **2007**, *111*, 4641–4644. [CrossRef] [PubMed]
22. Hayes, R.; Warr, G.G.; Atkin, R. Structure and Nanostructure in Ionic Liquids. *Chem. Rev.* **2015**, *115*, 6357–6426. [CrossRef] [PubMed]

23. Kondrat, S.; Bier, M.; Harnau, L. Phase behavior of ionic liquid crystals. *J. Chem. Phys.* **2010**, *132*, 184901. [CrossRef]

24. Saielli, G.; Bagno, A.; Wang, Y. Insights on the isotropic-to-Smectic a transition in ionic liquid crystals from coarse-grained molecular dynamics simulations: The role of microphase segregation. *J. Phys. Chem. B* **2015**, *119*, 3829–3836. [CrossRef] [PubMed]

25. Saielli, G.; Wang, Y. Role of the electrostatic interactions in the stabilization of ionic liquid crystals: Insights from coarse-grained MD simulations of an imidazolium model. *J. Phys. Chem. B* **2016**, *120*, 9152–9160. [CrossRef] [PubMed]

26. Kofu, M.; Tyagi, M.; Inamura, Y.; Miyazaki, K.; Yamamuro, O. Quasielastic neutron scattering studies on glass-forming ionic liquids with imidazolium cations. *J. Chem. Phys.* **2015**, *143*, 234502. [CrossRef] [PubMed]

27. Nemoto, F.; Kofu, M.; Yamamuro, O. Thermal and structural studies of imidazolium-based ionic liquids with and without liquid-crystalline phases: The origin of nanostructure. *J. Phys. Chem. B* **2015**, *119*, 5028–5034. [CrossRef] [PubMed]

28. Nemoto, F.; Kofu, M.; Nagao, M.; Ohishi, K.; Takata, S.; Suzuki, J.; Yamada, T.; Shibata, K.; Ueki, T.; Kitazawa, Y.; Watanabe, M.; Yamamuro, O. Neutron scattering studies on short- and long-range layer structures and related dynamics in imidazolium-based ionic liquids. *J. Chem. Phys.* **2018**, *149*, 54502. [CrossRef] [PubMed]

29. Nelyubina, Y.V.; Shaplov, A.S.; Lozinskaya, E.I.; Buzin, M.I.; Vygodskii, Y.S. A New volume-based approach for predicting thermophysical behavior of ionic liquids and ionic liquid crystals. *J. Am. Chem. Soc.* **2016**, *138*, 10076–10079. [CrossRef] [PubMed]

30. Quevillon, M.J.; Whitmer, J.K. Charge transport and phase behavior of imidazolium-based ionic liquid crystals from fully atomistic simulations. *Materials* **2018**, *11*, 64.

31. Peng, H.; Kubo, M.; Shiba, H. Molecular dynamics study of mesophase transitions upon annealing of imidazolium-based ionic liquids with long-alkyl chains. *Phys. Chem. Chem. Phys.* **2018**, *20*, 9796–9805. [CrossRef] [PubMed]

32. Bates, M.A.; Luckhurst, G.R. Computer simulation studies of anisotropic systems. XXX. The phase behavior and structure of a Gay-Berne mesogen. *J. Chem. Phys.* **1999**, *110*, 7087–7108. [CrossRef]

33. Margola, T.; Saielli, G.; Satoh, K. MD simulations of mixtures of charged Gay-Berne and Lennard-Jones particles as models of ionic liquid crystals. *Mol. Cryst. Liq. Cryst.* **2017**, *649*, 50–58. [CrossRef]

34. Saielli, G.; Margola, T.; Satoh, K. Tuning Coulombic interactions to stabilize nematic and smectic ionic liquid crystal phases in mixtures of charged soft ellipsoids and spheres. *Soft Matter* **2017**, *13*, 5204–5213. [CrossRef] [PubMed]

35. Plimpton, S. Fast parallel algorithms for short-range molecular dynamics. *J. Comput. Phys.* **1995**, *117*, 1–19. [CrossRef]

36. Wang, X.; Sternberg, M.; Kohler, F.T.U.; Melcher, B.U.; Wasserscheid, P.; Meyer, K. Long-alkyl-chain-derivatized imidazolium salts and ionic liquid crystals with tailor-made properties. *RSC Adv.* **2014**, *4*, 12476–12481. [CrossRef]

37. Rohini, R.; Lee, C.-K.; Lu, J.-T.; Lin, I.J.B. Symmetrical 1, 3-dialkylimidazolium based ionic liquid crystals. *J. Chin. Chem. Soc.* **2013**, *60*, 745–754. [CrossRef]

38. Bhowmik, P.K.; Han, H.S.; Akhter, S.; Han, H.S. Lyotropic liquid-crystalline main-chain viologen polymers. *J. Polym. Sci. Part A-Polymer Chem.* **1995**, *33*, 1745–1749. [CrossRef]

39. Bhowmik, P.K.; Han, H.S.; Cebe, J.J.; Burchett, R.A.; Acharya, B.; Kumar, S. Ambient temperature thermotropic liquid crystalline viologen bis(triflimide) salts. *Liq. Cryst.* **2003**, *30*, 1433–1440. [CrossRef]

40. Causin, V.; Saielli, G. Effect of a structural modification of the bipyridinium core on the phase behaviour of viologen-based bistriflimide salts. *J. Mol. Liq.* **2009**, *145*, 41–47. [CrossRef]

41. Berardi, R.; Emerson, A.P.J.; Zannoni, C. Monte Carlo investigations of a Gay-Berne liquid crystal. *J. Chem. Soc. Faraday Trans.* **1993**, *89*, 4069–4078. [CrossRef]

42. Gay, J.G.; Berne, B.J. Modification of the overlap potential to mimic a linear site-site potential. *J. Chem. Phys.* **1981**, *74*, 3316–3319. [CrossRef]

43. Berardi, R.; Fava, C.; Zannoni, C. A Gay–Berne potential for dissimilar biaxial particles. *Chem. Phys. Lett.* **1998**, *297*, 8–14. [CrossRef]

44. Brown, W.M.; Petersen, M.K.; Plimpton, S.J.; Grest, G.S. Liquid crystal nanodroplets in solution. *J. Chem. Phys.* **2009**, *130*, 44901. [CrossRef] [PubMed]

45. Hockney, R.; Eastwood, J. *Computer Simulation Using Particles*; Adam Hilger; CRC Press: New York, NY, USA, 1989.
46. Shinoda, W.; Shiga, M.; Mikami, M. Rapid estimation of elastic constants by molecular dynamics simulation under constant stress. *Phys. Rev. B* **2004**, *69*, 134103. [CrossRef]
47. Gabriel, A.T.; Meyer, T.; Germano, G. Molecular graphics of convex body fluids. *J. Chem. Theory Comput.* **2008**, *4*, 468–476. [CrossRef] [PubMed]
48. Frenkel, D.; Louis, A.A. Phase separation in binary hard-core mixtures: An exact result. *Phys. Rev. Lett.* **1992**, *68*, 3363–3365. [CrossRef] [PubMed]
49. Antypov, D.; Cleaver, D.J. The role of attractive interactions in rod–sphere mixtures. *J. Chem. Phys.* **2004**, *120*, 10307–10316. [CrossRef] [PubMed]
50. Wysocki, A.; Löwen, H. Oscillatory driven colloidal binary mixtures: Axial segregation versus laning. *Phys. Rev. E* **2009**, *79*, 41408. [CrossRef] [PubMed]
51. Vissers, T.; Van Blaaderen, A.; Imhof, A. Band formation in mixtures of oppositely charged colloids driven by an ac electric field. *Phys. Rev. Lett.* **2011**, *106*, 228303. [CrossRef] [PubMed]
52. Liu, F.; Chen, B.; Glettner, B.; Prehm, M.; Das, M.K.; Baumeister, U.; Zeng, X.; Ungar, G.; Tschierske, C. The trapezoidal cylinder phase: A new mode of self-assembly in liquid-crystalline soft matter. *J. Am. Chem. Soc.* **2008**, *130*, 9666–9667. [CrossRef] [PubMed]
53. Chen, B.; Zeng, X.; Baumeister, U.; Ungar, G.; Tschierske, C. liquid crystalline networks composed of pentagonal, square, and triangular cylinders. *Science* **2005**, *307*, 96–99. [CrossRef] [PubMed]
54. Bader, K.; Neidhardt, M.M.; Wöhrle, T.; Forschner, R.; Baro, A.; Giesselmann, F.; Laschat, S. Amino acid/crown ether hybrid materials: How charge affects liquid crystalline self-assembly. *Soft Matter* **2017**, *13*, 8379–8391. [CrossRef] [PubMed]
55. Kirchner, B.; Malberg, F.; Firaha, D.S.; Holl, O. Ion pairing in ionic liquids. *J. Phys. Condens. Matter* **2015**, *27*, 463002. [CrossRef] [PubMed]

MDPI

St. Alban-Anlage 66

4052 Basel

Switzerland

Tel. +41 61 683 77 34

Fax +41 61 302 89 18

www.mdpi.com

Crystals Editorial Office

E-mail: crystals@mdpi.com

www.mdpi.com/journal/crystals

www.ingramcontent.com/pod-product-compliance
Lightning Source LLC
Chambersburg PA
CBHW051915210326
41597CB00033B/6149